Molecular Genetic Mechanisms in Development and Aging

ACADEMIC PRESS RAPID MANUSCRIPT REPRODUCTION

*Proceedings of a Symposium
Held in Miami, Florida
March 24-25, 1972*

Molecular Genetic Mechanisms in Development and Aging

Edited by

Morris Rockstein
George T. Baker, III

Department of Physiology and Biophysics
University of Miami School of Medicine
Miami, Florida

Academic Press New York and London 1972

ACADEMIC PRESS, INC.
111 Fifth Avenue, New York, New York 10003

United Kingdom Edition published by
ACADEMIC PRESS, INC. (LONDON) LTD.
24/28 Oval Road, London NW1

LIBRARY OF CONGRESS CATALOG CARD NUMBER: 72-88346

PRINTED IN THE UNITED STATES OF AMERICA

CONTENTS

CONTRIBUTORS

M. B. Baird, Masonic Medical Research Laboratory, Utica, New York 13501

Martin Blumenfeld,* Department of Zoology, The University of Texas, Austin, Texas 78712

P. S. Chen, Institute of Zoology, University of Zurich, Zurich, Switzerland

Ulrich Clever, Department of Biological Sciences, Purdue University, Lafayette, Indiana 47907

Calvin A. Lang, Department of Biochemistry and the Biological Aging Program, University of Louisville School of Medicine, Louisville, Kentucky 40208

H. R. Massie, Masonic Medical Research Laboratory, Utica, New York 13501

Ruth R. Painter, Department of Environmental Toxicology, University of California, Davis, California 95616

Narayan G. Patel, Central Research Department, Experimental Station, E. I. du Pont de Nemours and Co., Wilmington, Delaware 19898

Morris Rockstein, Department of Physiology and Biophysics, University of Miami School of Medicine, Miami, Florida 33152

H. V. Samis, Masonic Medical Research Laboratory, Utica, New York 13501

James D. Stidham, Department of Biology, Presbyterian College, Clinton, South Carolina 29325

---

*Present Address: Department of Genetics, University of Wisconsin, Madison, Wisconsin 53706

PREFACE

In this publication, the culmination of a two-day symposium held in Miami from March 24 to 25, 1972, on Molecular Genetic Mechanisms in Development and Aging, we have been able to bring together a group of experts concerned with interpreting and explaining the mechanisms of aging at the level of the genome.

While much of our fundamental knowledge concerning the expression and regulation of genes is derived from studies on microbial organisms, the field of genetics, per se, has come full circle in the employment of insects as an organism of choice in the search for the molecular genetic basis for the processes of aging. Indeed, the highly complex multicellular members of the class Insecta provide the investigator with many unique and distinct advantages for aging studies. These include short life-spans, essentially post-mitotic tissues as adults, availability of highly inbred and genetically well-defined strains, and ease of handling and maintenance of large populations, to mention only a few.

To the participants in this symposium, we must extend our special thanks for their cooperativeness in making available so promptly, the manuscripts covering their presentations. To Dr. Bennett Sallman and Dr. George E. Schaiberger of the Department of Microbiology must go acknowledgment for their having served as Co-Chairmen during this two-day symposium. Special thanks and sincerest appreciation are due Mrs. Estella Cooney for her untiring efforts in arranging for this symposium from its inception as well as for typing the camera copy for this publication and for related editorial activities.

This work was supported in part by the University of Miami, Department of Physiology and Biophysics, as well as by the National Institute of Child Health and Human Development (Training Program in Cellular Aging—Grant No. HD-00142) without whose support this meeting could not have been held.

Morris Rockstein
George T. Baker, III

Molecular Genetic Mechanisms in Development and Aging

THE ROLE OF MOLECULAR GENETIC MECHANISMS
IN THE AGING PROCESS

Morris Rockstein, Ph.D.

Department of Physiology and Biophysics
University of Miami School of Medicine
Miami, Florida

For a considerable number of years, I have been
concerned with studying, together with my students
and colleagues, the underlying biochemical events
which precede, accompany, and also follow the maturation and decline of the locomotor ability of flight
in the male house fly, Musca domestica, L., with age.
Because the flight muscle in the male house fly shows
marked degenerative changes both in structure and
function in relation to declining flight ability, I
have been particularly interested in those biochemical components of flight muscle of flying insects
which are involved in the energizing of contraction
of the highly active flight muscle. In this regard,
we have found a persistent, reproducible pattern of
age-related sequential biochemical events, as shown
in Table I and in Figures 1 and 2. These changes
occur generation after generation according to a pre-
dictable and, therefore, assumedly genetically programmed series of step-wise biochemical alterations
from day to day after adult emergence.

This begins with a steep to slow rise in activity of each of four enzyme systems to a peak reached
on successive days following emergence of the adult
male fly (see Fig. 3), as follows:

1) The transphosphorylating enzyme arginine
phosphokinase (ATP:L-arginine phosphotransferase
EC 2.7.3.3), in both crude and purified forms, shows
a rapid rise in activity to a peak level at exactly

2 days after adult male emergence, following which there is a rapid decline for several days, followed by a steady decline to a gradual leveling off thereafter (Baker and Rockstein, unpublished). These age-dependent changes in enzymatic activity were not associated with detectable chemical, kinetic, or electrophoretic properties of the enzyme protein (Baker, 1971). This suggests that these are actual changes with age either in the rate of synthesis or in the degradation of this enzyme, rather than changes in content of co-enzymes, activators, or inhibitors, which might alter the overall activity of this enzyme with advancing age.

2) The extramitochondrial enzyme NAD-dependent alpha-glycerophosphate dehydrogenase (EC 1.1.1.99.5) (Rockstein and Brandt, 1963), important in the overall synthesis of ATP through the reoxidation of reduced NAD, shows a less steep rise from emergence to the peak at 4 days post-emergence, falling rather rapidly for the next several days and then more slowly to the third week of adult male existence, during which most of the males will have lost their wings through gradual abrading.

In a later study (Rockstein and Farrell, 1972), id٫ ١tical absolute mobilities and a homogeneous **enzy**me protein were obtained for alpha-glycerophosphate dehydrogenase on gel electrophoresis at all ages examined. Similarly, identical absolute mobilities and a homogeneous enzyme protein were found for arginine phosphokinase throughout the observed age-related changes in enzyme activities (Baker, 1971). Therefore, the age-related changes cannot be attributed to any variation in respective isozyme content as such.

3) The intramitochondrial enzyme Mg-activated ATPase rises to a peak at the 6th day of adult life which is coincident with the onset of failure of flight, especially marked in the initiation of the rapid rate of loss of wings in the adult male population (see Fig. 1), at this particular age.

4) The very gradual rise in the intramitochon-

drial cytochrome c oxidase activity to a peak at about the 10th to 11th day of adult life occurs coincident with the observed peak in the number of giant mitochondria (Rockstein and Bhatnagar, 1965). This is followed by a steady decline in parallel fashion of both the number of mitochondria as well as in enzyme activity.

5) As shown in Fig. 2, thiamine, one of the most widely required vitamins among multicellular animals, and the precursor of at least one important co-enzyme concerned with intermediary metabolism of carbohydrates, rises from emergence to a peak level at the 4th day of adult life. This increase is coincident with that of alpha-glycerophosphate dehydrogenase activity, after which both decline steadily (Rockstein and Hawkins, 1970).

6) The arginine phosphate content of the flight muscle, the source of the inorganic phosphate in the reconstitution of ATP by the enzyme arginine phosphokinase, rises steadily, in the male, to a peak 5 days following emergence, and falls steadily thereafter (Rockstein, unpublished). Finally, the substrate ATP is maintained at a low level in the thoracic flight muscle through the first 4 days post-emergence coincident with the rising level of the intramitochondrial Mg-activated ATPase (Rockstein and Gutfreund, 1961). It then begins to rise inversely, with the onset of decline of the aforementioned enzyme, at between the 5th and 6th days following emergence. Then, within 48 hours following the onset of decline in Mg-activated ATPase content, i.e., at 8 days of age, the ATP content of the thorax has undergone a dramatic increase to its peak level, which level remains comparatively unchanged (declining very slightly and very slowly) throughout the remainder of the life of the aging male house fly. Moreover, this peak, occurring as it does at the 7th and 8th day, occurs virtually at the same age as the precipitous onset of wing loss (see Fig. 1).

We are well aware of the widely held hypothesis that control of cellular function is encoded in the

DNA, in turn mediated through the DNA-directed synthesis of ribonucleic acid, which itself in turn provides the regulatory mechanisms for orderly protein synthesis. Accordingly, the control of such a programmed sequence of aging events as we observe in the male house fly flight muscle (in relation to the aging of locomotor ability of flight) must be sought within this highly complex molecular genetic nexus, now that the _facts_ of aging have been so established in this particular manifestation of senescence.

Presented with such a series of interrelated programmed biochemical, physiological and structural events, ultimately resulting in the total failure of flight, one is forced to infer that aging of locomotor ability in this species must be a genetically controlled and directed series of events.

According to that theory of aging which would attribute those events characteristic of "aging", cell damage would be mediated either by somatic mutation (Curtis, 1963), or through accumulated random errors occurring at any point within the molecular genetic chain of events, from transcription to the final steps of translation (Szilard, 1959). One might equally argue that the somatic mutation or the random cell damage theory may be applicable even in this case if one were to assume that the errors or accidents which result in genetic damage (von Hahn, 1966, 1970; Kurtz and Sinex, 1967) are occurring on a statistically distributional basis for any one species of animal exposed to an average, if not controlled, physical and biotic environment. On the other hand, one might suggest that such errors are not random but a built-in eventuality of the entire process of development and differentiation. However that may be, we are still led to the hypothesis that aging (particularly senescence) is related to alterations in the biosynthetic processes, especially with regard to important enzyme systems, the failure of which has been manifest in our studies on the house fly. Thus, the control of such alterations may occur in any step of this sequence of transmission of genetic information from DNA to the final steps of translation and, ultimately, the syntheses of proteins. Various aspects in the regulation of and alterations in this sequence will be discussed in other species during this two-day Symposium.

Accordingly, this Symposium, which I have been fortunate in being able to organize with the collaboration of Dr. Baker, under the sponsorship of the Training Program in Cellular Aging of the University of Miami, supported by the National Institutes of Child Health and Human Development, may serve to extend our understanding of the fundamental processes involved in development, maturation and senescence.

And now, to listen to as well as to discuss along with our audience the various aspects of the molecular genetic mechanisms concerned in aging and development, which our fine panel of experts will be examining during this two-day Symposium.

ACKNOWLEDGEMENT

The original research by Rockstein was supported in part by funds from the U.S. P.H.S. National Institutes of Child Health and Human Development Research Grant No. HD00571 and N.I.H. Research Grant No. GM 09680.

The original research by Baker was supported in part by funds from NICHD Research Grant No. HD00571 and Training Program Grant No. HD00142.

REFERENCES

Babers, F.H., and Pratt, J.J.,Jr. (1950). _Physiol. Zool._ _23_, 58.

Baker, G.T. (1971). Doctoral dissertation, University of Miami, Florida.

Baker, G.T., and Rockstein, M. (1972). (Unpublished).

Clark, A.M., and Rockstein, M. (1964). _In_ "Physiology of Insecta" (M. Rockstein, ed.), Vol. I, pp.259-281, Academic Press, New York.

Curtis, H.J. (1963). _Science_ _141_, 686.

Hahn, H.P. von (1966). J. Gerontol. 21, 291.
Hahn, H.P. von (1970). Gerontologia 16, 116.
Kurtz, D.I., and Sinex, F.M. (1967). Biochim. Biophys.
    Acta 145, 840.
Rockstein, M. (1967). Soc. Exptl. Biol. Symp. XXI,
    337.
Rockstein, M. (1972). (Unpublished).
Rockstein, M., and Bhatnagar, P.L. (1965). J. Insect
    Physiol. 11, 481.
Rockstein, M., and Bhatnagar, P.L. (1966). Biol. Bull.
    131, 479.
Rockstein, M., and Brandt, K.F. (1963). Science 139,
    1049.
Rockstein, M., and Farrell, G.J. (1972). J. Insect
    Physiol. 18, 737.
Rockstein, M., and Gutfreund, D.E. (1961). Science
    133, 1476.
Rockstein, M., and Hawkins, W.B. (1970). Experimental
    Gerontol. 5, 187.
Rockstein, M., and Srivastava, P.N. (1967). Experi-
    entia 23, 636.
Szilard, L. (1959). Proc. Nat. Acad. Sci. U.S. 45,
    30.

TABLE I

TIME SEQUENCE OF AGING OF
FLIGHT ABILITY IN THE MALE HOUSE FLY

Biological Parameter	Maximum	Reference
Acid phosphatase	emergence*	Clark and Rockstein, 1964
Trehalose content	4 hours	Rockstein and Srivastava, 1967
Brain cholinesterase	1 day	Babers and Pratt, 1950
Duration of flight	1 day	Rockstein and Bhatnagar, 1966
Arginine phosphokinase	2 days	Baker and Rockstein, 1972
Wing beat frequency	4-9 days	Rockstein and Bhatnagar, 1966
Thiamine content	4 days	Rockstein and Hawkins, 1970
Alpha-GDH	4 days	Rockstein and Brandt, 1963
Arginine phosphate	5 days	Rockstein, 1972
Alkaline phosphatase	5 days	Clark and Rockstein, 1964
Per cent wing loss	6 days#	Rockstein and Brandt, 1963
Mg-ATPase	6 days	Rockstein and Brandt, 1963
ATP content	8 days	Rockstein and Gutfreund, 1961
Number of mitochondria	8-12 days	Rockstein and Bhatnagar, 1965
Cytochrome $\underline{c}$ oxidase	11 days	Rockstein, 1967

* minimum at 5 days
# onset

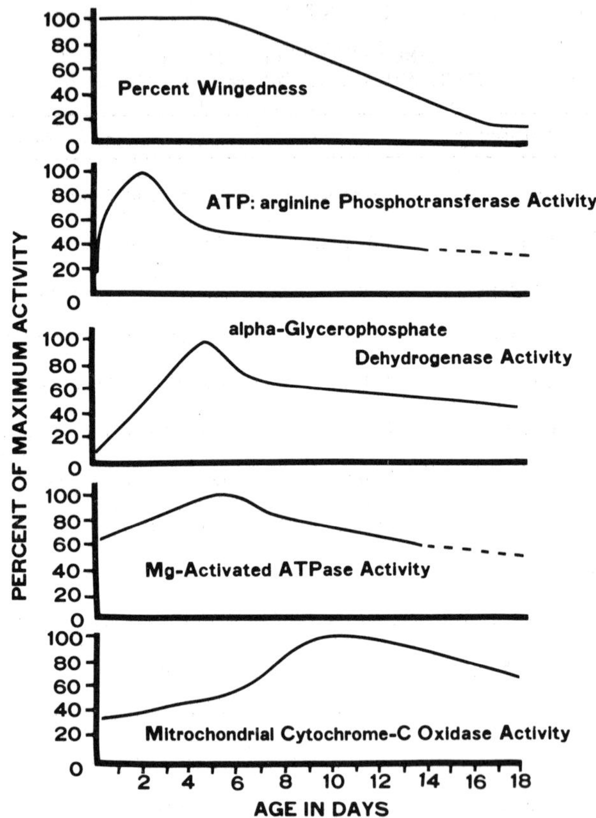

Fig. 1 Sequence of age-dependent changes in the male house fly.

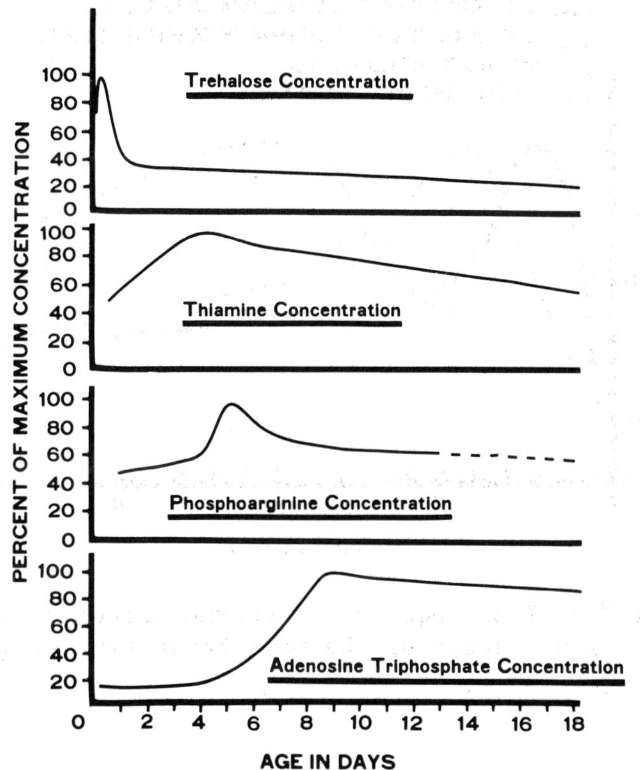

Fig. 2 Sequence of substrate concentration changes in the thorax of the male house fly, _Musca domestica_ L.

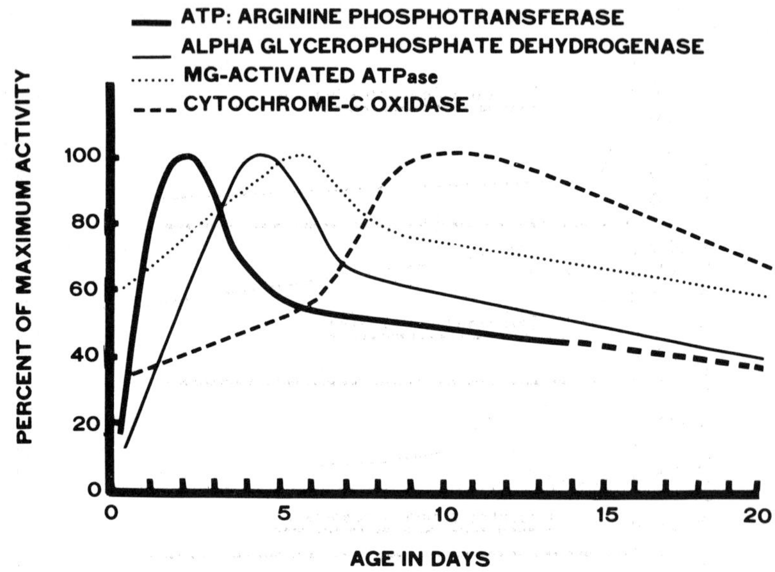

Fig. 3   Time sequence of enzyme activity changes in the flight muscle of the male house fly, _Musca domestica_ L.

THE REPLICATION OF SATELLITE DNA's
DURING <u>DROSOPHILA</u> DEVELOPMENT

Martin Blumenfeld*

Department of Zoology
The University of Texas
Austin, Texas 78712

The mechanisms of genetic control and genetic
organization during the development and aging of
eukaryotes currently pose some of the most challeng-
ing problems in cellular and developmental biology.
They include, for instance, the possible roles of
repeated DNA sequences, histones, RNA polymerases,
hormones, etc. in the processes of control and organization. This presentation is concerned with one of
these problems--the function of satellite DNA's.

I. Repeated sequences and satellite DNA's

Repeated nucleotide sequences, represented 10^3-
10^6 times per genome (Britten and Kohne, 1968, 1969),
are universally present in eukaryotic DNA's, but
rarely present in prokaryotic DNA's. They commonly
represent 20-60% of the eukaryotic genome, and constitute the clearest distinction between the DNA's of
eukaryotes and prokaryotes. Their wide distribution
attests to their importance.
    Satellite DNA's constitute an unusual and impor-
tant class of nuclear repeated sequences, set apart
from the major portion of the eukaryotic genome by
their distinctly different base compositions, which
cause them either to produce band asymmetry or to

*Present Address: Department of Genetics, The
University of Wisconsin, Madison, Wisconsin 53706

form separate bands during ultracentrifugation in
CsCl or Cs_2SO_4. Mitochondrial or chloroplast DNA's
which can also appear as density components, are not
considered satellites (McCarthy, 1969).

The base compositions of different satellite
DNA's vary widely. The $d(A-T)_n$-like satellites detected in crabs of the genus Cancer and in Drosophila
melanogaster contain less than 5% guanine + cytosine
(G+C) (Sueoka, 1961; Swartz et al., 1962; Fansler
et al., 1970); mouse satellite DNA contains 34% G+C
(Walker et al., 1969); ribosomal RNA-coding satellites (rDNA) may exceed 60% G+C (reviewed by Birnsteil et al., 1971).

Satellites have been reported in members of
virtually every eukaryotic group and in certain
halophilic bacteria (reviewed by McCarthy, 1969;
Hearst and Botchan, 1970). Their relative genomic
representation is species-specific. For instance,
they represent 41% of the nucleotide sequences in
diploid cell DNA of Drosophila virilis (Gall et al.,
1971) but only 12-14% in D. melanogaster (Gall et al.,
1971; Blumenfeld and Forrest, 1971).

The number of satellites per genome and their
distribution throughout the karyotype are also
species-specific. Among the insects, Oncopeltus
fasciatus DNA contains no detectable satellites
(Lagowski et al., in preparation); D. hydei DNA contains one (Dickson et al., 1971); D. virilis and D.
melanogaster DNA's each contain three (Gall et al.,
1971; Blumenfeld and Forrest, 1971). Among the
mammals, mouse DNA contains one satellite (Kit, 1961;
Walker et al., 1969); guinea pig DNA contains two
(Walker et al., 1969); human DNA contains at least
five (Allen, personal communication). Satellites are
heavily concentrated in the centromeric heterochromatin (reviewed by Arrighi et al., 1970; Gall et al.,
1971; Yunis and Yasmineh, 1971), but in a bewildering
fashion. Mouse satellite DNA is present in the centromeric heterochromatin of every chromosome except
the Y (Jones, 1970; Pardue and Gall, 1970; Gall and
Pardue, 1971). In contrast, some human satellites

are localized in the heterochromatin of one or two
chromosome pairs (Jones and Corneo, 1971; Saunders
et al., 1972) while at least one other human satellite DNA is widely distributed throughout the genome
(Jones and Corneo, 1971).

Biochemical analysis has revealed the existence
of two functionally different classes of satellites--
informational and non-informational. Informational
satellites are complex sequences that provide temp-
lates for the synthesis of specific kinds of RNA
molecules, such as ribosomal RNA precursors (reviewed by Birnsteil et al., 1971; Brown et al., 1972),
and sea urchin histone messenger RNA (Kedes and Birnsteil, 1971). Non-informational satellites are short,
tandemly repeated nucleotide sequences (Walker et
al., 1969; Southern, 1970; Thomas, 1971). They are
thought not to carry genetic information in the
classical sense because they are not complementary
to RNA sequences transcribed in vivo (Flamm et al.,
1969) and are heavily concentrated in centromeric
heterochromatin, where few structural genes have
been mapped (reviewed by Arrighi, et al., 1970; Gall
et al., 1971; Yunis and Yasmineh, 1971). The
function(s) of these non-informational satellites
has yet to be demonstrated. Their repeated occurrence
and concentration in centromeric regions suggest that
they may be involved in various chromosome-recogni-
tion processes (Walker et al., 1969).

On special occasions, cells can alter the balance between repeated and non-repeated sequences in
the nucleus. These "unbalancing acts" are accompanied
by dramatic alterations in nuclear morphology.

1. Gene amplification

During the premeiotic S phase of oogenesis in
many eukaryotes, rDNA genes are selectively replicated and released as extra nucleoli (reviewed by
Brown and Dawid, 1968; Gall, 1969; Lima-de-Faria et
al., 1969; Bird and Birnsteil, 1971; Brown and Blackler, 1972). rDNA amplification also occurs during

Wolffian regeneration of the newt lens (Collins, 1972), and after hormone treatment of cultured human cells (Koch and Cruceanu, 1971).

2. Under-replication during polytenization

Satellites and other repeated sequences are fully represented in diploid cells from embryos or larval imaginal discs of <u>D</u>. <u>melanogaster</u>, <u>D</u>. <u>virilis</u>, and <u>D</u>. <u>hydei</u>. However, they are under-represented in polytene larval salivary gland cells (Gall <u>et al</u>., 1971; Hennig and Meer, 1971; Dickson <u>et al</u>., 1971). This under-representation results from the association of heterochromatin in the chromocenter, and the progressive under-replication of the chromocenter during larval development (Mulder <u>et al</u>., 1968; Rudkin, 1969). Similarly, when the protozoan <u>Stylonychia</u> transforms one of its micronuclei into a macronucleus, nuclear DNA content increases, and three satellite DNA's are under-replicated (Bostock and Prescott, 1972).

3. Chromatin diminution

During chromatin diminution in somatic cells of <u>Ascaris</u> <u>lumbricoides</u>, chromosomes fragment and portions of the genetic material are lost. At the completion of this process, the representation of repeated sequences is reduced by 60%, and particular classes of repeated sequences are quantitatively eliminated (Tobler <u>et al</u>., 1972).

Since specialized cells can be programmed to increase or decrease genomic levels of specific kinds of repeated sequences, and since these changes may offer a clue to the function of non-informational satellites, it is of fundamental importance to understand the mechanisms that trigger differential replication, and also the means by which the cells identify different classes of repeated sequences. The following experiments represent one way of approaching these problems.

II. <u>Drosophila</u> Satellite DNA's

A. <u>D</u>. <u>melanogaster</u> embryos

The use of <u>D</u>. <u>melanogaster</u> embryos has been the cornerstone of these experiments. They are particularly attractive to students of developmental and molecular genetics because:

1) Their genetics are well defined (Lindsley and Grell, 1968), and can be applied to specific developmental and molecular problems (Fristrom, 1970; Wright, 1970; Schneiderman and Bryant, 1971).

2) They can be produced in the large quantities needed for the study of molecular events. Embryos are mass-produced by continuously-fed young adult flies maintained in population cages, and then collected (Blumenfeld and Forrest, 1971). Their purity, as estimated under a dissecting microscope, exceeds 99%. Their viability, as measured by the percentage of wild-type embryos (Oregon R) that hatch into first instar larvae, is 95%. The "typical" cage, containing 25 g young adult flies (25 x 10^3) produces approximately 30 g embryos (1.5 x 10^6) during its one week career. Peak production is at day 3 or day 4 of cage life, when 25 g adults produce 5-7 g embryos per 24 hours (Figure 1). Thereafter, embryo production falls off rapidly.

3) Their development can be synchronized. When a two-hour collection is made in the afternoon, 90% of the embryos hatch between 20.5 and 23.5 hours later (Figure 2). Thus, the synchrony of development, first observed by Powsner (1935) with small numbers of embryos, is not lost when population size is increased by several orders of magnitude.

B. The genetic approach to satellite DNA's

The basic idea has been to introduce extra chromosomes into populations of embryos and look for changes in the relative amounts of specific satellite

DNA's. The Y chromosome was chosen because it was
the easiest chromosome to manipulate genetically.
Judd (personal communication) suggested the use of
compound X chromosomes, in which two homologous elements share one centromere; compound XY chromosomes,
in which two heterologous elements share one centromere; and Y chromosome fragments, in which specific
portions of the Y chromosome are lost. Attached X
($\widehat{XX}$); attached X,Y ($\widehat{XY}$); attached X, long arm of Y
($\widehat{XY}^L$); attached X, short arm of Y ($\widehat{XY}^S$); long arm of
Y (Y^L); short arm of Y (Y^S); and, attached short arm
of Y ($Y^S \cdot Y^S$) chromosomes were used to construct populations containing varying numbers of extra Y chromosomes or extra portions of Y chromosomes (Table 1).

The distribution of extra Y chromosomes or extra
Y chromosome portions varied in different popula-
tions. For instance, in the $\widehat{XXY}$, $\widehat{XYY}$ stock, both
males and females carried an extra Y chromosome; in
the $\widehat{XXY}^S$, $\widehat{XY}^L Y^S$ stock, only the females carried an
extra Y^S.

DNA was purified from fresh or frozen embryos
by a modified Marmur (1961) procedure (Blumenfeld
and Forrest, 1971). Embryo DNA's were analyzed by
CsCl equilibrium centrifugation and by thermal denaturation. These methods were chosen because they
resolve a mixture of DNA's into components on the
basis of their G+C content. For DNA's with G+C contents between 20 and 70%, the higher the percentage
G+C, the higher the buoyant density in CsCl (Sueoka
et al., 1959; Rolfe and Meselson, 1959). Similarly,
the temperature at which DNA denatures and shows a
hyperchromic increase is also a function of base
composition. The higher the percentage G+C, the
higher the temperature of denaturation (Marmur and
Doty, 1962).

    C.   d(A-T)$_n$ and the Y chromosome of D. melano-
        gaster

When Drosophila embryo DNA preparations (XX,XY
and $\widehat{XXY}$, $\widehat{XYY}$) were centrifuged to equilibrium in

CsCl (Figure 3), they were resolved into three components (Fansler _et al._, 1970; Blumenfeld and Forrest, 1971). These were a major component (ρ = 1.702), a less-dense satellite component (ρ =1.687), and an extremely light satellite (ρ =1.675) which is called _Drosophila_ $d(A-T)_n$ because its G+C content is less than 5% (Fansler _et al._, 1970). Both "control" and "extra Y" DNA's contained the same components in about the same relative proportions.

However, when the same DNA's were analyzed by thermal denaturation, a more informative picture emerged. Hyperchromicity analysis revealed that _D. melanogaster_ DNA denatures stepwise (Figure 4). The earliest denaturing component ($48-52^{\circ}C$) is _Drosophila_ $d(A-T)_n$ (Blumenfeld and Forrest, 1971). After $d(A-T)_n$ denatures, there is no further increase in hyperchromicity until the temperature reaches $58^{\circ}C$. This plateau permits a direct measurement of the percentage hyperchromicity due to the denaturation of $d(A-T)_n$, which can be equated with the percentage representation of $d(A-T)_n$. The validity of this approach has been confirmed by preparative ultra-centrifugation in $Cs_2 SO_4-HgCl_2$ gradients (Blumenfeld and Forrest, 1971).

Figure 4 also illustrates that $d(A-T)_n$ represents a higher percentage of the total hyperchromic increase in DNA from extra Y embryos than in DNA from wild-type embryos. Extra Y DNA contains 6% $d(A-T)_n$ while wild-type DNA contains 4% $d(A-T)_n$ (Table 2). The difference between these values is significant at the 99% confidence levels (Blumenfeld and Forrest, 1971).

Since $d(A-T)_n$ levels were increased by the addition of extra Y chromosomes, it was important to know whether this effect was associated with either the long or the short arm of the Y chromosome, or whether it required an entire, intact Y chromosome. These possibilities were tested by measuring $d(A-T)_n$ percentages in DNA purified from embryos containing extra Y^S or Y^L chromosome arms. Embryos carrying one extra Y^L, or two or more extra Y^S chromosome arms, have significantly higher $d(A-T)_n$ values than do control embryos (Table 3). Clearly, additional portions of the Y chromosome raise $d(A-T)_n$ levels.

The correlation between Y chromosome dosage and $d(A-T)_n$ concentration can be interpreted in two ways. 1) $d(A-T)_n$ is heavily concentrated on the Y chromosome; 2) the Y chromosome carries a factor that stimulates $d(A-T)_n$ synthesis or retards $d(A-T)_n$ degradation. The distinction between these alternatives should be testable by <u>in situ</u> hybridization experiments (Ball and Pardue, 1971).

D. The replication of satellite DNA's during <u>Drosophila</u> development

The polytene salivary gland cells of <u>D</u>. <u>melanogaster</u>, <u>D</u>. <u>virilis</u>, and <u>D</u>. <u>hydei</u> larvae have reduced representations of satellite DNA, reflecting the under-replication of heterochromatin during polytenization (Gall <u>et al</u>., 1971; Dickson <u>et al</u>., 1971). Since <u>Drosophila</u> pupae and adults also contain polytene cells (cf. Ashburner, 1970 for a review), as well as diploid cells, they might be expected to have levels of satellite DNA reduced in proportion to their polytene cell content. Therefore, experiments were undertaken to answer the following questions: Are satellite DNA sequences under-represented in pupae and adults, as well as in larvae? Are different satellites equally under-represented in adults?

1. The replication of <u>D</u>. <u>melanogaster</u> $d(A-T)_n$

The low levels of $d(A-T)_n$ in wild-type embryo DNA make differences between embryos, pupae, and adults difficult to quantitate. However, amplification of $d(A-T)_n$ by the introduction of extra Y chromosomes, facilitated the measurement of changes in the relative amounts of $d(A-T)_n$ during the development of XXY, XYY individuals (Blumenfeld and Forrest, 1972).

Genomic levels of $d(A-T)_n$ in developing $\hat{X}XY$, $\hat{XY}Y$ populations were determined by thermal denaturation. The percentage of $d(A-T)_n$ was measured in DNA's isolated from 0-4 hour embryos, 18-22 hour embryos, pupae, and adults. $d(A-T)_n$ constitutes 6% of the DNA in embryos, but only 3% of the DNA in pupae and adults (Table 4, Blumenfeld and Forrest, 1972).

The following conclusions can be drawn from these results: 1) virtually all embryonic nuclei are diploid (this being the simplest explanation of the constant $d(A-T)_n$ level during embryonic development); 2) the relative representation of $d(A-T)_n$ is reduced in pupae and adults. This 50% reduction in $d(A-T)_n$ levels can be attributed to the presence of polytene cells in pupae and adults. It suggests that 50% of the DNA from pupae or adults is contributed by polytene cells. If the average reduplication of non-repeated sequences in polytene cells is arbitrarily taken to be 128-fold, these cells constitute approximately 1% of the cell population in pupae and adults. This estimate is reasonable, when one considers that adults contain a variety of polytene cell types (cf. Ashburner, 1970).

2. Differential under-replication of satellite DNA in _D. virilis_

Since $d(A-T)_n$ is dramatically under-represented in adults, it became important to determine if other satellites were equally under-represented. While _D. melanogaster_ contains three satellites (Blumenfeld and Forrest, 1971), it was not ideally suited for this experiment because convenient, accurate determinations were possible only for $d(A-T)_n$. We therefore turned to _D. virilis_, in which the proportions of satellites are much higher (Gall _et al._, 1971), to test this possibility. The representation of satellite DNA's in _D. virilis_ embryo DNA and adult DNA preparations was compared by thermal denaturation and by analytical ultracentrifugation.

The thermal denaturation profiles of embryo DNA and adult DNA are similar. However, the T_m (the temperature at which 50% denaturation is attained (Marmur and Doty, 1962)) of embryo DNA is about $1.5^{\circ}C$ lower than that of adult DNA (Figure 5), suggesting that the base compositions of the two DNA's are different. Further information about this difference was obtained by densitometric tracings from CsCl equilibrium centrifugation of D. virilis DNA preparations (Figure 5). DNA preparations from both embryos and adults are resolved into a main band (I: ρ = 1.700) and three less-dense satellite bands (II: ρ = 1.691; III: ρ =1.688; IV: ρ =1.669). The buoyant density values are in agreement with those published by Gall et al., (1971). Bands II, III, and IV represent 23, 9, and 8%, respectively, of total embryo DNA, and 16, 4, and 8%, respectively, of total adult DNA. The representations of bands II and III are reduced by about 50% in adults, but the representation of band IV is not reduced. This result indicates that different DNA satellites are not equally under-represented in adults, and are therefore differentially under-replicated in the polytene cells of adult flies.

Band IV from D. virilis and $d(A-T)_n$ from D. melanogaster have similar buoyant densities in CsCl. Paradoxically, band IV is fully replicated in adults while $d(A-T)_n$ is under-replicated. Chemical analyses have not been done on band IV; therefore, it is difficult to reach a final verdict on the meaning of this apparent discrepancy. However, the following lines of evidence support the view that band IV and $d(A-T)_n$ are not the same, even though they have similar densities in CsCl.

1) D. virilis DNA does not denature appreciably in the temperature range (48-52°C) where D. melanogaster $d(A-T)_n$ denatures (compare Figures 4 and 5), indicating that D. virilis DNA lacks a component that corresponds to D. melanogaster $d(A-T)_n$.

2) The G+C content of very light satellites cannot be inferred from their densities in CsCl. For

instance, the light satellites present in DNA's of
two related crab species have the same density in
CsCl, but dramatically different base compositions
(Skinner _et al._, 1970).

III.   Speculations: How are repeated sequences
        recognized?

These results may contain a clue to the molecular mechanisms involved in repeated-sequence under-
replication and recognition. Since different density
satellites, which by definition possess different
base compositions, are differentially under-replicated during polytenization, then some feature of the
satellite DNA molecule, besides its repeatedness, is
involved in the recognition process. Since the con-
formation of DNA, as judged by X-ray scattering, is
a function of its base composition (Bram, 1971),
satellites and other repeated sequences may have
distinctive conformations, imposed by their base
compositions, which allow specific recognition by
one or more kinds of proteins. Conformational differences in repeated sequences would provide a simple
mechanism by which selected portions of the genome
could be recognized by specific proteins. After
recognition these regions could be selectively ex-
cluded from replication. Conformationally distinct
repeated sequences could also serve as regulatory
factors that control gene expression in eukaryotes,
i.e., as the sensor genes of the Britten-Davidson
(1969) model. Similar ideas have been proposed
recently (Bram, 1971; Crick, 1971).
The key question remains, "How do repeated sequences act?". It is almost certain that their action
involves recognition by one or more kinds of proteins. The experiments which I have presented
suggest that some feature of the repeated sequence,
besides its repeatedness, is involved in recognition
by specific proteins.

ACKNOWLEDGEMENT

These experiments were performed in collaboration with Hugh S. Forrest at The University of Texas, Austin. They were supported by grants from the National Institutes of Health (HD-03803 and GM-15-769 -05) and the Robert A. Welch Foundation, Houston, Texas. I wish to thank Y. Hiraizumi, Burke Judd, Jeanne Lagowski, and D. Pavan for their advice and criticism.

REFERENCES

Allen, J. (1972). (Personal communication.)

Arrighi, F.E., Hsu, T.C., Saunders, P., and Saunders, G.F. (1970). Chromosoma 32, 224.

Ashburner, M. (1970). Advanc. Insect Physiol. 7, 1.

Bird, A.P., and Birnsteil, M.L. (1971). Chromosoma 35, 300.

Birnsteil, M.L., Chipchase, M., and Speirs, J.(1971). Prog. Nucleic Acid Res. Mol. Biol. 11, 351.

Blumenfeld, M., and Forrest, H.S. (1971). Proc. Nat. Acad. Sci. U.S. 68, 3145.

Blumenfeld, M., and Forrest, H.S. (1972). (Submitted for publication.)

Bostock, C.J., and Prescott, D.M. (1972). Proc. Nat. Acad. Sci. U.S. 69, 139.

Bram, S. (1971). Nature New Biol. 232, 174.

Britten, R.J., and Davidson, E.H. (1969). Science 165, 349.

Britten, R.J., and Kohne, D.E. (1968). Science 161, 529.

Britten, R.J., and Kohne, D.E. (1969). In "Handbook of Molecular Cytology", (A. Lima-d-Faria, ed.) pp. 21-36. North Holland, Amsterdam.

Brown, D.D., and Blackler, A.W. (1972). J. Mol. Biol. 63, 75.

Brown, D.D., and Dawid, I.B. (1968). Science 160, 272.

Brown, D.D., Wensink, P.E., and Jordan, E. (1972). J. Mol. Biol. 63, 57.

Collins, J.M. (1972). Biochemistry 11, 1259.

Crick, F.H.C. (1971). Nature 234, 25.

Dickson, E., Boyd, J.B., and Laird, C.D. (1971). J. Mol. Biol. 61, 615.

Fansler, B.S., Traviglini, E.C., Loeb, L.A., and Schultz, J. (1970). Biochem. Biophys. Res. Comm. 40, 1266.

Flamm, W.G., Walker, P.M.B., and McCallum, M. (1969). J. Mol. Biol. 40, 423.

Fristrom, J. (1970). Ann. Rev. Genetics 4, 323.

Gall, J.G. (1969). Genetics Suppl. 61, 121.

Gall, J.G., Cohen, E.H., and Polan, M.L. (1971). Chromosoma 33, 319.

Gall, J.G., and Pardue, M.L. (1971). In "Methods in Enzymology", XXI, pp. 470-480 (L. Grossman and K. Moldave, eds.), Academic Press, New York.

Hearst, J.E., and Botchan, M. (1970). Ann. Rev. Biochem. 39, 151.

Hennig, W., and Meer, B. (1971). Nature New Biol. 233, 70.

Jones, K.W. (1970). Nature 225, 912.

Jones, K.W., and Corneo, G. (1971). Nature New Biol. 233, 268.

Judd, B. (1972). (Personal communication.)

Kedes, L.H., and Birnsteil, M.L. (1971). Nature New Biol. 230, 165.

Kit, S. (1961). J. Mol. Biol. 3, 711.

Koch, J., and Cruceanu, A. (1971). Hoppe-Seyler's Z. Physiol. Chem. 352, 137.

Lagowski, J.M., Laird, C.D., and Forrest, H.S. (In preparation.)

Lindsley, D.L., and Grell, E.H. (1968). "Genetic Variations of Drosophila melanogaster." Carnegie Institution of Washington, Publication No. 627, Washington, D.C.

Lima-de-Faria, A., Birnsteil, M., and Jaworska, H. (1969). Genetics Suppl. 61, 145.

Marmur, J. (1961). J. Mol. Biol. 3, 202.

Marmur, J., and Doty, P. (1962). J. Mol. Biol. 5, 109.

McCarthy, B.J. (1969). In "Handbook of Molecular Cytology", (A. Lima-de-Faria, ed.) pp. 3-20, North Holland, Amsterdam.

Mulder, M.P., Van Duijn, P., and Gloor, H.J. (1968). _Genetica_ 39, 385.

Pardue, M.L., and Gall, J.G. (1970). _Science_ 168, 1356.

Powsner, L. (1935). _Physiol. Zool._ 8, 274.

Rolfe, R., and Meselson, M. (1959). _Proc. Nat. Acad. Sci. U.S._ 45, 1039.

Rudkin, G.T. (1969). _Genetics Suppl._ 61, 227.

Saunders, G.F., Shirakawa, S., Saunders, P.P., Arrighi, F.E., and Hsu, T.C. (1972). _J. Mol. Biol._ 63, 323.

Schneiderman, H.A., and Bryant, P.J. (1971). _Nature_ 234, 187.

Skinner, D.M., Beattie, W.G., and Kerr, M.S. (1970). _Nature_ 227, 837.

Southern, E.M. (1970). _Nature_ 227, 794.

Sueoka, N. (1961). _J. Mol. Biol._ 3, 31.

Sueoka, N., Marmur, J., and Doty, P. (1959). _Nature_ 183, 1429.

Swartz, M.N., Trautner, J.A., and Kornberg, A. (1962). _J. Biol. Chem._ 237, 1961.

Tobler, H., Smith, K.D., and Ursprung, H. (1972). _Dev. Biol._ 27, 190.

Thomas, C.A., Jr. (1971). _Ann. Rev. Genetics_ 5, 425.

Walker, P.M.B., Flamm, W.G., and McLaren, A. (1969). _In_ "Handbook of Molecular Cytology", (A. Lima-de-Faria, ed.) pp. 52-66, North Holland, Amsterdam.

Wright, T.R.F. (1970). _Adv. Genetics_ 15, 262.

Yunis, J.J., and Yasmineh, W.G. (1971). _Science_ 174, 1200.

TABLE 1

Genetic Methods*

| Karyotype | | Extra Y Chromosomes or Y |
$\female\female$	$\male\male$	Chromosome Arms Per $\female$ + $\male$
1. XX	XY	0
2. $\hat{X}X/Y$	$\hat{X}Y/Y$	2 extra Y chromosomes
3. $\hat{X}X/Y^S$	$\hat{X}Y^L/Y^S$	1 extra Y^S chromosome arm
4. $\hat{X}Y^S$ XY^S	$\hat{X}Y^S/Y^L$	2 extra Y^S chromosome arms
5. $\hat{X}X/Y^S \cdot Y^S$	$\hat{X}Y^L/Y^S \cdot Y^S$	3 extra Y^S chromosome arms
6. $\hat{X}X/Y^L$	$\hat{X}Y^S/Y^L$	1 extra Y^L chromosome arm

*Adapted from Blumenfeld and Forrest (1971).

TABLE 2

The Effect of Extra Y Chromosomes on $d(A-T)_n$ Percentage*

Karyotype	% $d(A-T)_n$ ($\pm$ s. d.)
1. Wild Type (XX, XY)	3.8 $\pm$ 0.6 (4)
2. Extra Y ($\hat{X}XY$, $\hat{X}YY$)	6.3 $\pm$ 0.6 (3)**

** P $\ll$ 0.01

Each value represents the average $d(A-T)_n$ measurement for the indicated strain. The number of DNA samples analyzed for each strain is listed in parentheses. Samples within each group do not differ significantly from one another (F=1.09, n_1=5, n_2=17). However, the $\hat{X}XY$, $\hat{X}YY$ samples differ significantly from the XX, XY samples (F=92.2, n_1=1, n_2=22).

*From Blumenfeld and Forrest (1971).

TABLE 3

Effect of Extra Y^S or Y^L Chromosome
Arms on $d(A-T)_n$ Percentage*

Karyotype	$\% \ d(A-T)_n$ ($\pm$ s. d.)
1. Wild type (XX, XY)	3.8 ± 0.6 (4)
2. 1 Extra Y^S ($\hat{XX}/Y^S$, $\hat{XY}^L/Y^S$)	3.9 ± 0.4 (1)
3. 2 Extra Y^S ($\hat{XY}^S \ \hat{XY}^S$, $\hat{XY}^S/Y^L$)	5.4 ± 0.6 (3) **,a
4. 3 Extra Y^S ($\hat{XX}/Y^S \cdot Y^S$, $\hat{XY}^L/Y^S \cdot Y^S$)	6.5 ± 0.6 (2) **,b
5. 1 Extra Y^L ($\hat{XX}/Y^L$, $\hat{XY}^S/Y^L$)	6.3 ± 0.4 (2) **,c

**$P \ll 0.01$

Each value represents the average $d(A-T)_n$
measurement for the indicated strain. The number of
DNA samples analyzed for each strain is listed in
parentheses. Samples within each group do not differ
significantly from one another. (a. $F = 2.40$, $n_1 = 18$;
b. $F = 1.29$, $n_1 = 4$, $n_2 = 17$; c. $F = 0.79$, $n_1 = 4$, $n_2 = 14$).
However, $\hat{XY}^S \ \hat{XY}^S$, $\hat{XY}^S/Y^L$; $\hat{XX}/Y^S \cdot Y^S$, $\hat{XY}^L/Y^S \cdot Y^S$; and
$\hat{XX}/Y^L$, $\hat{XY}^S/Y^L$ samples differ significantly from the
XX, XY samples. (a. $F=38.2$, $n_1=1$, $n_2=23$; b. $F=73.7$,
$n_1=1$, $n_2=21$; c. $F=81.5$, $n_1=1$, $n_2=18$).

*From Blumenfeld and Forrest (1971).

TABLE 4

Genomic Levels of $d(A\text{-}T)_n$ During
the Development of $X\hat{X}Y, X\hat{Y}Y$ Flies*

Stage of Development	% $d(A\text{-}T)_n$ ($\pm$ s. d.)
0-4 hour embryos	6.0 ± 0.5 (2)
18-22 hour embryos	6.1 ± 0.4 (2)
pupa	3.0 ± 0.7 (2)
adult	2.9 ± 0.9 (3)

The number of different DNA preparations
analyzed for each stage is listed in parentheses.
The differences between and within groups were
analyzed by the F-test.

Samples within each group do not differ significantly from one another ($F=2.73$, $n_1=5$, $n_2=15$; $0.10 > P > 0.05$). Values for 0-4 hour embryos do not differ significantly from those for 18-22 hour embryos; values for pupae do not differ significantly from those for adults ($F=0.11$, $n_1=2$, $n_2=15$, $P > 0.10$). Values for embryos differ significantly from those for pupae or adults ($F=106.58$, $n_1=1$, $n_2=15$, $P \ll 0.01$).

*From Blumenfeld and Forrest (1972).

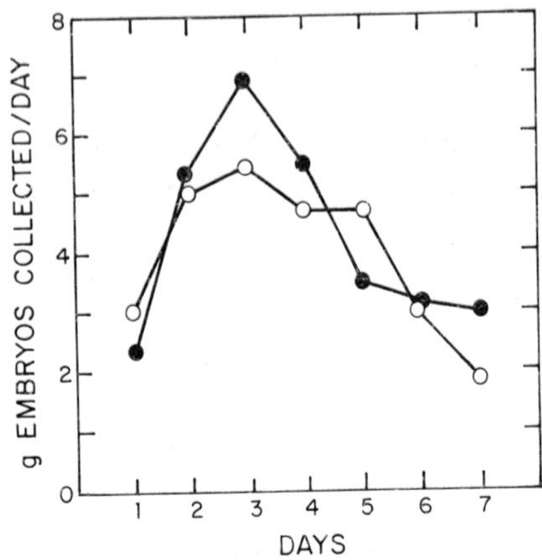

Fig. 1 The embryo-producing careers of two typical D. melanogaster population cages, each containing 25 g wild-type (Oregon R) adults. Embryos were collected daily, dried, and weighed.

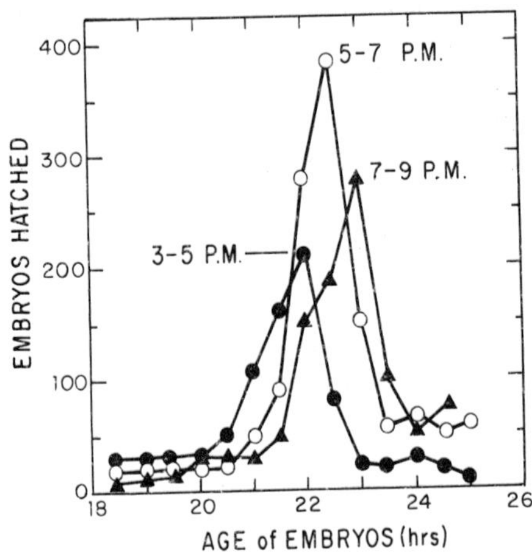

Fig. 2 The relative synchrony of D. melanogaster (Oregon R) embryos collected over three different two hour periods. The number of embryos hatching into first instar larvae was plotted as a function of age.

28

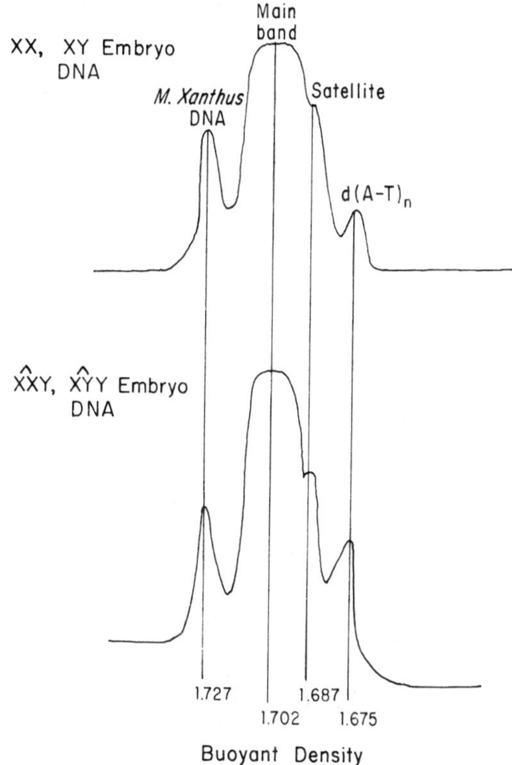

Fig. 3 Densitometric tracings following CsCl equilibrium centrifugation of D. melanogaster embryo DNA preparations (XX, XY and $\hat{X}$XY, $\hat{X}$YY). Myxococcus xanthus DNA (ρ =1.727) was used as an internal density marker. (From Blumenfeld and Forrest, 1971.)

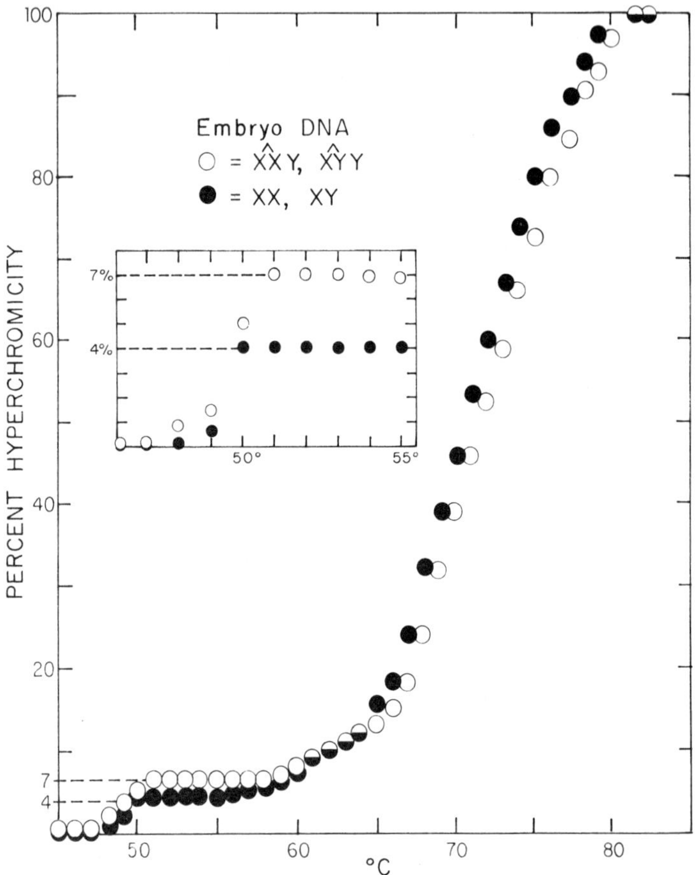

Fig. 4 Thermal denaturation of <u>D</u>. <u>melanogaster</u> embryo DNA preparations ((●) XX, XY and (O) X̂XY, X̂YY) in 0.015 M NaCl, 0.0015 M sodium citrate, pH 7.0. Insert, magnification of area around 50°C. (From Blumenfeld and Forrest, 1971.)

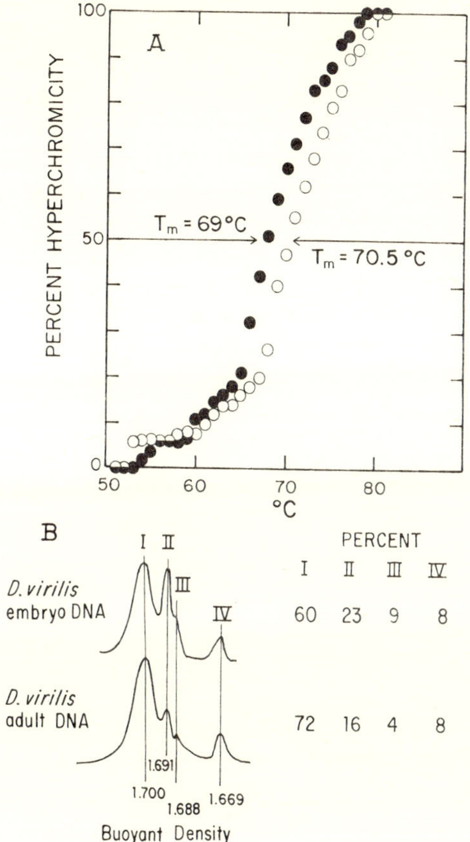

Fig. 5 Comparisons of D. virilis embryo and adult DNA preparations. A. Thermal denaturation of embryo (●) and adult (O) DNA's in 0.015 M NaCl, 0.0015 M sodium citrate, pH 7.0. (From Blumenfeld and Forrest, 1972.) B. Densitometric tracings following CsCl equilibrium centrifugation of embryo DNA (1.5 µg) and adult DNA (2.0 µg). Micrococcus luteus DNA (ρ =1.731) was used as an internal density standard. (From Blumenfeld and Forrest,1972.)

THE CONTROL OF CELLULAR GROWTH AND DEATH
IN THE DEVELOPMENT OF AN INSECT

Ulrich Clever

Department of Biological Sciences
Purdue University
Lafayette, Indiana 47907

The title of this essay may seem somewhat unusual in that the terms growth, aging and death refer to individual cells rather than to the organism as a whole. Cells, it would appear, participate in these events only indirectly: their growth usually is restricted to regaining the original size after cell division, and their death occurs when the organism as a whole dies. There are exceptions to these roles, however, of which insects provide classical examples. In the higher insects, a large portion of the larval cells is discarded at specific stages of metamorphosis, stages during which the remainder of the insect is actively developing. In fact, it appears to be the same signals, the molting hormones, which inform some cells to prepare for death whereas they inform other cells to replicate, to divide and to differentiate. Not only is death controlled on an individual-cell basis, but in many insects the same is true for growth. During larval development of these insects, cell division is quite rare and the tremendous growth that the larvae undergo is essentially based upon the selective growth of some of their cells. The genetic material of these cells multiplies many times during their growth, leading to the giant and highly polyploid cells characteristic of many insect larvae. In many cases the cells which have "opted" for growth during the larval stages are those destined for death once metamorphosis has begun.

The cells of the larval salivary glands of
Chironomus _tentans_ belong to this group. A pair of
these glands is formed during embryonic development,
each gland consisting of approximately 40 to 45
cells. This number does not change until the glands
degenerate in the young pupa, although the glands
grow dramatically. The chromatids in these cells do
not separate following endomitosis but remain paired
to a multistranded cable, the polytene chromosomes.
Changes in chromosomal activity, as visualized by
puffing, can thus be studied in these cells and
attempts can be made to correlate some of these
changes to cell growth and death. I shall discuss,
in this article, aspects of the growth of the salivary gland cells during the last larval instar, and
of their preparation for death early in metamorphosis.

A. Cell Growth

Chironomus larvae need about 3 weeks to reach
the last larval instar, and an additional 3 to 4
weeks to reach the adult stage. Larvae which have
just completed ecdysis from the 3rd to the last (4th)
larval instar are distinguished from older larvae by
the bright red appearance of their heads, resulting
from the absence of black pigments in the still un-
sclerotized cuticle. This stage (the red-head or
RH-stage) lasts about 24 hours. The age of last instar larvae will be given here in days after the RH-
stage. We shall consider two parameters of cell
growth during the last larva instar: the change of
protein content and the change of genome size.
The gland of a RH-larva contains approximately
4 µg of protein. The protein content increases to
approximately 17 µg in older last instar larvae
(Fig. 1). However, the total glandular protein consists of secretory proteins, which are contained in
the glandular lumen, in addition to cellular proteins. The contribution of the secretory proteins
has been estimated by allowing them to flow out of

explanted glands after microinjury (Grossbach, 1969) and by treatment of larvae with a solution of pilo-carpin which causes the release of the secretory proteins (Grossbach, 1969; Darrow and Clever, 1970). Both methods yield similar values for the proportion of secretory protein in the gland. While there is considerable variation among larvae, the average is approximately 40% throughout the intermolt period of the last instar (Darrow and Clever, 1970). There-fore, as indicated in Fig. 1, cellular protein in-creases during the last instar from approximately 2.5 µg to approximately 10 µg. Since there are ap-proximately 40 to 45 cells per gland, this means that on the average each cell during the last instar grows from roughly 0.06 µg protein at the RH-stage to 0.23 µg protein at day 20. Although this estimate is based on the simplified assumption that all cells of the salivary gland are of approximately equal size, which is not quite correct, it does provide a rough idea about the size of the cells we are deal-ing with.

The amount of DNA in a diploid nucleus of <u>Chironomus</u> <u>tentans</u> is approximately 0.45 pg, or 2.6 x 10^{11} daltons, measured by Feulgen spectrophotometry (Daneholt and Edström, 1967) or estimated from re-association kinetics of denatured DNA (Sachs and Clever, 1972). The largest nuclei of salivary gland cells of last instar larvae contain 3360 µg of DNA, corresponding to 13 duplication steps during polyten-ization. The same number of duplication steps has been estimated from nuclear volume measurements, assuming that this volume doubles with each replica-tion step. The volume of a diploid nucleus is ap-proximately $32\mu^3$ and that of the largest salivary gland cell nuclei approximately $500,000\mu^3$ (Beermann, 1952). The same method was used to estimate the average number of replications per nucleus during the last larval instar (Fig. 2). Nuclei in the glands of RH-larvae have replicated on the average 9.8 times, whereas nuclei of 20-day-old larvae have replicated 12.4 times. Thus, during the 3 weeks of the last instar the nuclei replicate 2 or 3 times. The variation in size between nuclei of individual glands may be seen from the inset of Fig. 2 to range over 3 to 4 nuclear size classes. The distribution of the larger and the smaller nuclei (and cells) is not random. For example, the cells close to the secretory duct of the gland are usually small. Nothing is known, however, about the determination of these differences in cell size, in itself an interesting problem.

Synthetic activity is obviously required for the growth in cellular protein and DNA during the last instar. This, however, is only part of the total cellular synthetic activity during this time, the remainder serving the secretory function of the gland. Like other secretory cell types, the gland cells show a well developed rough endoplasmic reticulum (Fig. 4a; see also Kloetzel and Laufer, 1969) and only a very few, if any, active polyribosomes not attached to membranes (Clever and Storbeck, 1970). Few quantitative data of the glandular secretory activity are available. From autoradiographic observations, using pulses with tritiated lysine, Doyle and Laufer (1969a) estimate that about 80% of the newly synthesized proteins are released into the lumen during a 4 hour chase period. This may be an over-estimate since the secretory proteins are exceptionally rich in basic amino acids (Grossbach, 1969), but it does serve to indicate that a major portion of the glandular activity is continuously devoted to its secretory functions. The portion of glandular protein contained in the secretory lumen rises sharply during the early part of the prepupal period, i.e., the period during which the animal prepares for the larval-pupal ecdysis (Fig. 1). It is not known whether this is due to an increased rate of synthesis of secretory proteins, or to a decline in their release.

Throughout the last larval instar, puffs and Balbiani rings in the glands' polytene chromosomes are actively engaged in RNA synthesis. It would appear, thus, that the synthetic activities described

in the preceding paragraphs might be under immediate
genomic control, i.e., continuously monitored by newly synthesized informational RNA. However, it also
appears that glandular protein synthesis can continue
unaffected for very long periods of time without concurrent RNA synthesis. This was shown for total
glandular protein synthesis by Clever et al. (1969),
who kept larvae for up to two days in actinomycin
solutions and then pulsed for 30 minutes with labeled
amino acids. There was no marked difference between
the actinomycin-treated larvae and their untreated
controls. Similar results were also shown for secretory proteins (Doyle and Laufer, 1969b) and for the
production of the special secretory granules in Ch.
pallidivittatus (Clever, 1969), originally shown by
Beermann (1961) to be synthesized under the control
of a Balbiani ring locus. Likewise, if polyribosomes
are extracted and fractionated in sucrose gradients,
there is no difference in the profiles obtained from
glands of control larvae and of those kept in actinomycin (Fig. 3a; Clever and Storbeck, 1970). Thus,
it would appear that a large fraction of mRNA in
these cells is very stable and does not need to be
continuously replaced.

Protein synthesis is largely independent of concurrent RNA synthesis throughout the last larval instar except for larvae of the RH-stage (Clever et al.,
1969; Clever and Storbeck, 1970). For example, more
polyribosomes are recovered from glands of one-day-
old larvae than from those of RH-larvae, suggesting
an increase in protein synthesis. If, however, the
RH-larvae are kept in actinomycin during this day,
the amount of polyribosomes recovered not only does
not increase, but actually falls below that of the
original RH-larvae (Fig. 3b). Some of the mRNA in
glands of RH-larvae, thus, appears to be relatively
short-lived and needs to be continuously replaced.
In addition, the apparent increase in protein synthesis during the first day of the last instar also
seems to depend upon concurrently synthesized RNA.
Evidence indicating that some specific RNA involved

in growth processes is synthesized at the RH-stage
comes from a study of the control of DNA replication.
The rate of replication, expressed as the frequency of nuclei incorporating ^{3}H-thymidine during a
30-minute incubation, is highest during the intermolt stages, in the 3rd as well as in the last larval
instar (Fig. 5; Darrow and Clever, 1970). On the
other hand, during the larval molt from the 3rd to
the last instar and the larval-pupal molt there is
little or no replication. Similar changes in replication frequency are shown by the Malpighian tubules
(Fig. 5). It is clear that these ceasements of replication, like all molt-related processes, are ultimately under the control of the molting hormones.
Rodman (1968), who made similar observations in
Drosophila salivary glands, suggests that an "anti-
initiator" is formed in response to high levels of
ecdysone, preventing the initiation of new replication cycles. However, in Chironomus, replication
continued unaffected for about 2 days when a pupal
molt was induced precociously with high dosages of
β-ecdysone. This would seem to suggest that it is
not the hormone concentration per se which controls
the decline of replication, although the possibility
of a rapid inactivation of the injected ecdysone
must be kept in mind in our experiments (cf., Robbins
et al., 1971).
From the data of Figs. 2 and 5 the average
length of the S-period in the salivary gland nuclei
of last instar larvae may be estimated to be between
10 and 20 hours. If the time needed for replication
increases with the degree of polyteny (Keyl and
Pelling, 1963), then the S-period would, on the
average, be correspondingly shorter in the earlier
part of the last intermolt stage when the replication
frequency is highest. When RNA or protein synthesis
is blocked for 48 hours in larvae 3 days old or older,
the rate of replication is not or, at most, minimally
affected (Table 1). Since this inhibition period is
much longer than the S-phase, these findings show
that in larvae of these stages, neither the continua-

tion of DNA synthesis nor its initiation depend upon concurrent or immediately preceding synthesis of RNA or of protein. The latter conclusion could be avoided only if there were a considerable lengthening of the S-phase caused by the drug treatments, which our observations make unlikely (Darrow and Clever, 1970). The apparent independence of replication from concurrent RNA and protein syntheses in polytenic cells seems in contrast to findings in diploid cell lines. Conceivably, this is related to the fact that succeeding replication cycles are not separated by cell divisions in the polytenic cells.

In contrast to these findings with mid-last instar larvae, the resumption of replication following the larval molt is blocked by inhibition of RNA synthesis as well as of protein synthesis (Table 1). One could explain this stage-specific requirement of RNA and protein synthesis by postulating that entirely different mechanisms control replication at the two stages - not a very attractive assumption. We prefer as a working hypothesis that some protein of the replication machinery is synthesized only at the beginning of the instar, and that this protein is stable and remains operative in some or all of the subsequent replication cycles. Its synthesis requires the concurrent synthesis of RNA. The few RH-larvae which escape the drug effects may be those in which the required macromolecules already had been produced. This is supported by the fact that older RH-larvae are less sensitive than younger ones. It is tempting to speculate that the proteins required for replication at the RH-stage are among those synthesized from the unstable mRNA characteristic of this stage. It is well known that specialized cells may be programmed by a characteristic population of stable mRNA. Our results suggest that, in addition, cells may become programmed by stable proteins, which are synthesized from unstable mRNA at definite stages of development.

The uniqueness of the RH-stage does not seem to be restricted to developmental processes of the

salivary gland cells. Larvae placed into actinomycin at the RH-stage for only one day will invari-
ably die a few days later, although RNA synthesis
readily resumes when the larvae are changed into a
drug-free medium. Larvae which are placed for one
day into actinomycin at day 1, on the other hand,
continue to develop and finally pupate and metamor-
phose.

The observations discussed so far could be explained by postulating quantitative as well as qualitative changes in gene activity from the RH-stage
to the later stages. Since we are dealing with polytenic cells, qualitative changes in the pattern of
active genes might have been expected to be detectable by changes in the pattern of puffs. However,
the pattern of puffing in RH-larvae is very similar
to that in older larvae. Differences seem to be
restricted to some puffs which had been active during
the preceding molting period and which terminate
their activity only during the RH-stage (Clever,
1962a, 1963). Since the sensitivity of protein synthesis to an inhibition of RNA synthesis seems to
begin after ecdysis, it is unlikely that these
"carry-over" puffs are responsible for it (Clever et
al., 1969). While it is possible that some differences concerning small puffs were overlooked in these
studies, alternative interpretations of the apparent
contradiction are possible. We shall return to this
problem when discussing the metabolism of RNA at the
RH-stage and at later stages. For the later stages
10-day-old larvae were used as examples.

When glands from 10-day-old larvae are incubated
with ^{3}H-uridine, incorporation proceeds approximately
linearly for a short period of time and then declines
(Rubinstein and Clever, 1972; Fig. 6). As was shown
by appropriate controls, the decline is not caused by
deteriorating conditions of the glands during the
incubation period. Several pieces of circumstantial
evidence also argue strongly against the possibility
that the decline is due to changes in size or specific activity of the cellular UTP pools. For example,

total radioactivity in UTP reaches a plateau rapidly and remains at this plateau for the period of our experiments; there is no change in TCA-soluble absorbance at 260 nm during incubation; pool expansion is, under these circumstances, inconsistent with preincubation data; and, in addition, _de novo_ synthesis of pyrimidine nucleotides seems to be inhibited when uridine is present in the medium (cf., Plagemann, 1971). Thus, the kinetics of uridine incorporation shown for glands of 10-day-old larvae in Fig. 6 are best explained by assuming some of the RNA synthesized to be relatively short-lived. This interpretation finds support in the results of two additional, independent experiments. First, when RNA is extracted and fractionated by polyacrylamide gel electrophoresis, the fraction of label recovered in ribosomal RNA increases relative to that in heterodisperse RNA (> 15S) with the length of the pulse (Rubinstein and Clever, 1972). We assume that the ratio of the synthetic rate of rRNA to that of heterodisperse RNA does not change in favor of rRNA during our incubations, and that there is only a single precursor pool of the two species of RNA. It follows then, that heterodisperse RNA in glands of 10-day-old larvae is less stable than ribosomal RNA. Second, if the glands are pulsed with ^{3}H-uridine for 15 minutes and then transferred into a chase medium containing actinomycin D, at a concentration sufficient to block RNA synthesis instantly, about 50% of the incorporated radioactivity becomes acid-soluble within 15 minutes (Fig. 7). The RNA which becomes acid soluble during the actinomycin chase is heterodisperse RNA of very large molecular weight (cf., Rubinstein and Clever, 1972).

In contrast to these results with 10-day-old larvae, incorporation of ^{3}H-uridine into RNA of glands from RH-larvae proceeds linearly for several hours and does not decline (Fig. 6). The cellular UTP pools behave in a similar fashion in RH- as in 10-day-old larvae, i.e., they rapidly reach equilibrium with the externally supplied precursor and their

specific activity seems to remain constant thereafter. The possibility that in RH-larvae, but not
in 10-day-old larvae (or in both, but at different
rates) some of the externally supplied ^{3}H-uridine is
converted to cytidine nucleotides of RNA was also
ruled out. That this conversion does not occur at
either stage in <u>Chironomus</u> salivary glands has been
demonstrated (Rubinstein and Clever, 1972). From
the incorporation kinetics, we are left with the
conclusion, then, that in glands from RH-larvae the
majority of newly synthesized RNA is of considerably
longer life span than in glands from 10-day-old
larvae. This conclusion is supported by the fact
that (i) in RH-glands, in contrast to the older
glands, newly synthesized ribosomal RNA species do
not accumulate with time of incubation relative to
heterodisperse RNA; and (ii) that there is very
little, if any, degradation of newly synthesized RNA
discernible in RH-glands during chase experiments
with actinomycin D (Fig. 7).

Heterodisperse, high molecular weight RNA which
is degraded shortly after its synthesis, presumably
without leaving the nucleus, is a characteristic
feature of many animal cells (cf., Scherrer and
Marcaud, 1968; Darnell <u>et al</u>., 1970). Evidence that
such RNA exists in the polytenic salivary gland of
<u>Chironomus</u> <u>tentans</u> has also been reported by Daneholt
<u>et al</u>. (1969) and by Daneholt and Svedhem (1971).
The fact that such very short-lived RNA exists in
polytenic cells requires a re-evaluation of the
possibility that the RNA accumulation characteristic
of puffing might result from differential rates of
RNA turnover rather than from differential rates of
synthesis. This does not seem to be generally true,
however, since even at the RH-stage, when most newly
synthesized RNA is stable, puffs and Balbiani rings
are the preferentially labeled loci in autoradiographs (Rubinstein and Clever, 1972). Nevertheless,
the possibility exists that some of the variations
in puff sizes, or even some puff inductions, may
result from local changes in RNA degradation rather

than from changes in the rate of RNA synthesis. In
support of this possibility, there are indications
that artifactually induced puffs in Drosophila may
result from a transient stabilization of short-lived
RNA (Ellgaard and Clever, 1971).

The dramatic difference in the proportion of
rapidly degraded RNA between developmental stages,
as observed in <u>Chironomus</u>, is a novel observation.
It is intriguing to speculate that it is related to
the differences in the requirement of newly synthesized RNA for protein synthesis. For example, in
10-day-old larvae, where little such RNA is required,
a large portion of it is rapidly degraded, whereas
in RH-larvae, where newly synthesized RNA is required, there is no or little degradation. Consistent
with this, in the prepupae, where actinomycin also
interferes with protein synthesis (Clever <u>et al</u>.,
1969), a larger portion of newly synthesized RNA
is more stable than in 10-day-old intermolt larvae
(Fig. 7).

The function of the rapidly degraded nuclear
RNA in eukaryotic cells is still a matter of specu-
lation. It has been proposed that there may be a
precursor-product relationship between this RNA and
cytoplasmic mRNA, implying the existence of a mechanism allowing selective processing of the heavy
nuclear RNA into mRNA (cf., Scherrer and Marcaud,
1968; Scherrer <u>et al</u>., 1970; Darnell <u>et al</u>., 1970).
The correlation observed in <u>Chironomus</u> would be consistent with this model. One might even speculate
that the same RNA species are synthesized at the RH-
stage and in the older larvae and that the differences in protein synthesis are controlled posttran-
scriptionally by varying the population of RNA molecules released into the cytoplasm. This, of course,
would explain the puzzling observation reported
above, that the puffing patterns do not seem to
change from the RH-stage to the mid-instar stages.
Daneholt and Svedhem (1971) reported evidence suggesting that mRNA coding for secretory proteins is
the predominant type of RNA selectively stabilized

in older mid-instar larvae. At the RH-stage, additional species of mRNA might be stabilized which are
involved in the growth processes of the glands and
in programming the gland for the beginning instar.
In order to test the validity of the concept out-
lined here, the RNA species synthesized at the different stages will have to be compared qualitatively,
possibly by DNA-RNA hybridization experiments. Experiments of this type may be aided by the fact that
in <u>Chironomus</u>, in contrast to many other organisms,
more than 95% of the DNA seems to be of the single-
copy type (Sachs and Clever, 1972).

B. <u>Cell Death</u>

Cellular growth in the salivary glands gradually
ceases as the larvae mature (Fig. 1). The gland
cells respond to the hormonal condition initiating
metamorphisis by preparing for cytolysis. In mature
larvae and young prepupae replication is restricted
to very few (randomly distributed) nuclei, and it
comes to a complete halt in mid- to old prepupae
(Fig. 5). The glands release their luminar content
shortly before larval-pupal ecdysis. Following this,
cell degeneration begins. Some cellular protein is
already lost between the old prepupal and the young
pupal stage (Fig. 1), and the beginning destruction
of cellular ultrastructure is apparent in glands of
young pupae (Fig. 4b). In most pupae the glands
have entirely disappeared about 10 hours following
larval-pupal ecdysis. Glandular destruction is accomplished by autolytic processes. Whether the remnants of the gland cells are eventually removed by
macrophages, as it has recently been described for
<u>Drosophila</u> <u>pseudoobscura</u> (Harrod and Kastritsis,
1972), has not yet been studied. Based on morphological criteria and the puffing pattern, young prepupae, mid-prepupae (mpp) and three stages of old
prepupae (opp-1, opp-2 and opp-3) are distinguished
in <u>Ch</u>. <u>tentans</u> (Clever, 1962a; Henrikson and Clever,
1972).

In contrast to the situation discussed above
for the transition from the RH-stage to the later
intermolt stages, the transition from the intermolt
stage to the prepupal stage is accompanied by extensive changes in the puffing pattern (Fig. 8; Clever,
1962a). There is one group of puffs which newly
appear or become larger during the prepupal period
and another group of puffs which become inactive
during this time. As may be seen from Fig. 8, most
of the puffs which had been active during the intermolt stage, become gradually inactive during the prepupal period. In old prepupae these puffs are found
only in occasional animals and then they are very
small. This decline in puff activity may reflect the
gradual decline in functional activity of the gland
as it prepares for regression. Some of the products
of larval glands are no longer found in glands of
prepupae, such as, for example, a secretory protease
with a pH optimum of 5.5 (Rodems _et al_., 1969).
Enzymes in the larval glands of _Ch_. _thummi_ which disappear during the prepupal stage include malate dehydrogenase, trehalase, hyaluronidase and an esterase
(Laufer, 1968).

A number of other puffs appear or enlarge during
the prepupal period, puffs $11\text{-}B_2$, 14-B and 18-C of
Fig. 8 may be mentioned as examples. Puffing at loci
like these would be consistent with the idea that
changes in genomic activity are involved in the control of the developmental processes leading to cell
breakdown. We will turn to some of these processes
now and examine their requirement of genomic activity. We will concentrate our discussion on the synthesis, maturation and changing localization of two
enzymes which are presumably involved in cell destruction, acid phosphatase and an acid protease.

The glandular acid protease has a relatively
sharp pH optimum in homogenates or sonicates at 3.5 -
4.0. (Rodems _et al_., 1969; Henrikson and Clever,
1972). The characteristics of this enzyme, its low
pH optimum, strictly intracellular localization (see
below) and catalytic properties, closely resemble

those of cathepsin D, a lysosomal enzyme well known
for its involvement in autolytic processes. The pH
optimum of the glandular acid phosphatase was reported by Laufer and Schin (1971) to be 4.2; additional
catalytic properties of this glandular enzyme have
not yet been published.

In cytochemical preparations of glands from
larvae or prepupae, the acid phosphatase reaction
product is restricted to two types of membrane-
bounded bodies, small secretory granules and larger,
more electron-lucid structures (Schin and Clever,
1965, 1968a; Fig. 4a). The latter ones predominate
in the glands of prepupae. We will loosely refer to
them as "lysosome-like". It appears that the acid
phosphatase-containing cell organelles in the salivary gland arise from the Golgi apparatus, as has
been similarly suggested for "primary lysosomes"
(Cohn and Fedorko, 1969). The intracellular localization of the pH 3.5 protease has not yet been
studied cytochemically. However, differential centrifugation experiments reveal that at least 94% of
the enzyme, and presumably all of it, is particle
bound (Table 2). The enzyme-containing particles
show a buoyant density in sucrose gradients of 1.17
to 1.21 g/cm^3, identical with that of the acid phosphatase-containing lysosomes (Henrikson and Clever,
1972; see also Bowers et al., 1967). Since all lysosome-like structures showed acid phosphatase activity
in our previous study, we assume that both enzymes
are contained in the same cell organelles. Additional
evidence in support of this conclusion will be presented below.

Acid phosphatase activity is already present in
glands of 3rd instar larvae (Schin and Clever, 1968a)
and presumably in younger larvae as well. The activity increases throughout the last larval instar and
reaches a maximum at the time of gland regression
in young pupae (Laufer and Schin, 1971). A low level
of proteolytic activity also is present throughout
the last larval instar (Fig. 9). During the intermolt period, enzyme activity increases only slightly,

approximately in correspondence with the growth of
the gland. However, following the initiation of the
pupal molt, activity of the pH 3.5 protease increases
sharply and reaches a maximum in prepupae of stage
opp-3. As may be seen from the curve giving protease activity per glandular protein (specific activity) in Fig. 9 and from a comparison of Figs. 1 and
9, the increase in protease activity does not follow
the change in secretory proteins contained in the
glandular lumen. This and the fact that no enzyme
at any stage is released from the gland by pilocarpin
treatment shows that the enzyme is located entirely
intracellularly and is not part of the secretory product. Even when the total glandular protease activity no longer rises, the specific activity of the
enzyme continues to increase from stage opp-3 to the
young pupal stage (Fig. 9), indicating that there is
a preferential retention of the pH 3.5 protease during tissue degeneration (see Fig. 1). The same seems
to be true for acid phosphatase (Laufer and Schin,
1971). Thus, their intracellular localization, and
the course of their increase in activity during development strongly suggest that these enzymes participate in the processes of gland destruction. Their
prior existence in glands of intermolt larvae is consistent with the existence of some functionally
active lysosomes at this stage (Schin and Clever,
1968b).

In the present context, the rise of enzyme
activity during the prepupal stage and its control
is of particular interest. _A priori_, it might be due
to an increase in the rate of enzyme synthesis, to a
decrease in the rate of enzyme turnover, or to enzyme
activation at the level of existing and previously
synthesized enzyme molecules. In the former case,
the increase of enzyme activity during the prepupal
stage should be blocked by inhibitors of protein synthesis. Cycloheximide is the inhibitor of choice in
Chironomus since it is very effective when simply
added to the larval cultures. Glandular protein synthesis is inhibited very fast and, depending upon the

dose, to more than 99%.

Addition of cycloheximide to the larval culture medium does not by itself influence glandular pH 3.5 protease activity. This can be clearly seen from the fact that at stages where there is not normally a change in enzyme activity, there is no change of enzyme activity in the glands of larvae kept in cycloheximide. This is true for larvae with their low enzyme activity as well as for pupae with their high activity (Fig. 10). It is obvious from these data, however, that the enzyme molecules are very stable, i.e., their turnover rate is very low, at least under the experimental conditions employed. In prepupae kept in cycloheximide, the activity of the pH 3.5 protease continues to increase. The extent of this increase is specific for each stage (Fig. 10). In old prepupae of stage opp-2, enzyme activity increases during the first 24 hours in cycloheximide up to the level characteristic of opp-3 prepupae and pupae, thereafter, continued exposure to cycloheximide up until 48 hours resulted in no further increase in enzyme activity. In stage opp-1 prepupae, on the other hand, enzyme activity increases during the first 24 hours in cycloheximide up to the level of opp-2 prepupae, and continues to increase up to the pupal level during the next 24 hours. In mid-pre-pupae (mpp), finally, enzyme activity increases during a 24 hour stay in cycloheximide to the level of opp-1 prepupae, but then the increase comes to a halt.

These observations leave no doubt that the increase in enzyme activity during the later portion of the prepupal period is not due to the synthesis of new enzyme molecules at this time. Instead, it must result from the activation of existing enzyme molecules by a process not requiring the concurrent synthesis of protein. The results provide some indication as to the time of synthesis of the enzyme protein. We will return to this question after we have discussed the activation process.

Results of experiments in which gland homogen-

ates from larvae of various developmental stages, or
fractions of such homogenates, were mixed before
assaying for enzyme activity argue against the possibility that the enzyme activation is due to the presence of activators or inhibitors at the stages of
high or low enzyme activity, respectively. The involvement of metal ions in the activation process
also was ruled out (Henrikson and Clever, 1972).
Enzyme activation is achieved, however, by treatment
of gland homogenates with low concentrations of
trypsin.

The effectiveness of trypsin-treatment, just as
the increase of enzyme activity in cycloheximide, is
restricted to glands of those stages in which the
pH 3.5 protease activity is rapidly increasing in
normal development (Table 3). Thus, there is a 40 to
60% increase of enzyme activity following trypsin
treatment in prepupae of stages mpp and opp-1. Pupae
and old prepupae of stage opp-3, with their high
enzyme activity, and intermolt larvae with their low
activity, on the other hand, show little or no effect. The possibility that trypsin acts by releas-
ing enzyme from its particle association following
insufficient homogenization is ruled out, since
Triton X-100 has no comparable effect, and since
trypsin has no effect in opp-3 prepupae when all
enzyme is still particle bound. It would appear,
therefore, that trypsin treatment simulates the activation step of normal development which is independent of protein synthesis, possibly by removing resi-
dues from an inactive zymogen. Limited proteolysis
is a well-known mechanism for zymogen activation of
various protolytic enzymes, such as trypsinogen and
chymotrypsinogen, and has also been described for
several enzymes in insects (Ohnishi et al., 1970;
Berger et al., 1971). It appears to be in the form
of such a zymogen that the pH 3.5 protease is synthesized.

As was described above, the active pH 3.5 protease of prepupae is contained in lysosome-like particles. It was of obvious interest to see where the

zymogen is located in the cell and where its activation might proceed. For this reason, lysosome and supernatant fractions were prepared from prepupae of stages mpp and opp-3 and treated with trypsin. Enzyme activity in the lysosome fraction of mpp - prepupae was greatly enhanced, whereas there was no effect on either supernatant fraction or on the lysosomal fraction obtained from glands of opp-3 prepupae (Table 3). Thus, the zymogen is sequestered in lysosome-like particles shortly after its synthesis and it is here where it is activated. At the opp-3 stage the intralysosomal activation process is completed and is consistent with the data presented in Fig. 10.

The prepupal increase in enzyme activity, and the destruction of the gland, are events characteristic of metamorphosis and have ultimately to be under the control of the molting hormone, ecdysone. In fact, Radford and Misch (1971) recently reported that the number of lysosomes and the activity of a lysosomal enzyme (acid phosphatase) in the gut of the flesh fly, _Sarcophaga_, increases following an injection of β-ecdysone. When a pupal molt is induced precociously in larvae of _Chironomus_ by injection of ecdysone (cf., Clever, 1961), proteolytic enzyme activity does not change during the first two days, although the larvae reach the early mpp-stage (Table 3). However, at this time there is a slight but reproducible activability by trypsin, suggesting that the accumulation of inactive zymogen has begun (Table 3). As described above, when mpp-prepupae are placed into cycloheximide, enzyme activation proceeds, but it does so only to a limited extent (Fig. 10), consistent with the idea that the zymogen is being synthesized at this stage. In prepupae of the opp-1 stage, on the other hand, enzyme activity reaches its maximal level in the absence of protein synthesis (Fig. 10), indicating that zymogen synthesis is now complete. Thus, it would appear that the enzyme protein responsible for the high proteolytic activity at cytolysis is synthesized in an inactive

form in mid-prepupae. Since the effect of ecdysone
on zymogen accumulation is prevented by actinomycin
(Table 3), it would also appear that the synthesis
of this protein requires the concurrent or shortly
preceding synthesis of RNA. However, tempting as
they may be, these conclusions have to be considered
tentative at the moment. The use of immunochemical
methods would be required to verify them by determining the actual amounts of enzyme protein at any given
stage. Unfortunately, this is experimentally not
manageable at the present time.

Although the activity of the pH 3.5 protease
reaches its maximum late in the prepupal stage, cell
destruction is evident only after larval-pupal ecdysis. Furthermore, in opp-3 prepupae as well as in
very young pupae, cell destruction can still be prevented by blocking protein synthesis; it was mentioned above (Fig. 10) that the activity of the pH 3.5
protease is not affected. Thus, a high level of proteolytic enzyme activity may be a necessary, but it
is not a sufficient condition for the beginning of
cell destruction. The release of lysosomal enzymes
to their cellular substrates may be an additional requirement.

In very old prepupae and especially in young
pupae, lysosomes are filled with remnants of various
cell organelles, such as mitochondria, endoplasmic
reticulum and myelin-like structures (Fig. 4b). Thus,
it appears that cytolysis initially proceeds in
"autophagic vacuoles". In the glands of prepupae
and in relatively undegraded cells or areas of cells
in pupal glands, acid phosphatase is restricted to
the autophagic vacuoles. On the other hand, in cells
or cellular areas of pupal glands in which cell regression is more advanced, lysosomes appear to be
broken and acid phosphatase is now freely distributed
in the degrading cytoplasm (Fig. 4b). If it is
correct, as we concluded above, that the pH 3.5 protease is sequestered in the same organelles as acid
phosphatase, one would expect that this enzyme, too,
is no longer entirely particle bound in pupal glands.

The data presented in Table 2 confirm this; in pupal
glands about 30% of the enzyme is not sedimentable.
Since an inhibition of protein synthesis prevents
gland regression even at very late stages, it was of
interest to see whether it would also prevent the
liberation of the enzyme from its particle association. For this purpose, very old prepupae were
placed into cycloheximide for 24 hours. Most control
prepupae had pupated at the end of this time, and
their glands had entirely regressed or were in various stages of regression. In contrast, the glands
in all animals of the cycloheximide-treated group
appeared still visually intact. Less than 5% of
their proteolytic activity was non-sedimentable, in
contrast to the 30% normally found in young pupae.
Thus, whatever mechanism is causing the release of
enzyme from their lysosomal association, it seems to
involve the synthesis of some new protein. A similar
observation was reported by Lockshin (1969) for
muscle breakdown in the silkworm.

The final trigger that starts cell regression
in the glands, possibly by initiating the processes
leading to lysosome breakdown, is not yet known. However, the distinctiveness of the processes preparing
for and initiating cell breakdown and the requirement of a separate signal for the latter seems to be
a general phenomenon in insects. Lockshin and
Williams (1965a, b) presented evidence that the ces-
sation of motor-nerve impulses is the ultimate signal
which initiates histolysis of the intersegmental
muscles in silkmoths. Whether similar mechanisms are
at work in <u>Chironomus</u> remains to be seen.

C. Concluding Remarks

The observations discussed here lend support to
the notion that activity of a cell's genome is involved in the control of both its growth and its preparation for death. It is also clear, however, that
controls operating at the level of transcriptional
activity of the genome are only one part of a compli-

cated control network. Our results may provide a
wedge to gain some insight into this developmental
control network.

For example, it appears that the release of
genetic information from the nucleus itself is not
only controlled by variable patterns of gene activity, i.e., at the transcriptional level. Rather, a
mechanism of stabilization of newly synthesized RNA
seems to operate, possibly selecting the individual
RNA species to be released from an unchanging population of RNA synthesized. RNA species which are
concerned with the control of growth processes, such
as DNA replication, may be among those selected post-
transcriptionally. This level of information processing is now getting more and more attention and
the stage-specific differences in stabilization described here will provide a powerful tool for such
studies.

Secondly, it appears that cellular growth may
become programmed at particular development stages
by the synthesis of relatively stable proteins, such
as those which appear to be synthesized at the beginning of the new larval instar. While the synthesis
of these particular proteins may be unique to polytenic cells, in which succeeding replication cycles
are not interrupted by cell divisions, the programming of a cell for its subsequent development at the
level of stored proteins may be a more general phenomenon to which attention will have to be paid.

Thirdly, while the preparation for cell death
is initiated by an extracellular signal, ecdysone, it
is in itself a complicated intracellular process
which may be controlled at a multitude of succeeding
levels. If we take the pH 3.5 protease as an example
the first level of control might involve genomic
activity. The requirement of newly synthesized RNA
for the increased synthesis of this enzyme at a time
of characteristic changes in the puffing pattern may
indicate that transcriptional control mechanisms are
involved here. However, our data are also consistent
with the idea of selective mRNA stabilization as the

control mechanism involved. Subsequent levels at which control of cell regression might operate are the synthesis of the inactive enzyme precursor molecule, the sequestration of this precursor in lysosomes, the intralysosomal activation of the zymogen to the active enzyme, and finally the release of the sequestered enzyme to its cellular substrates. This release again is presumably triggered by extracellular factors. That cell death may be controlled at any of these levels during normal development is indicated by the observation that even in very old prepupae the preparation for cell death may come to a halt when these prepupae fall into a state of dormancy (Clever, 1962b). While this control network is obviously quite complex, its delineation opens an important aspect of insect development, programmed cell death, to our experimental analysis. It may be added that programmed cell death is, of course, an important part of morphogenetic processes in a variety of animals (cf., Saunders, 1966).

Finally, we are confronted with the ubiquitous question of developmental biology, why do these cells behave the way they do, that is, why is RNA metabolism at the RH-stage different from that at other stages, why do gland cells prepare for death in response to the same signal that causes other cells to grow and differentiate. Stage specificity of cellular behavior in insects is, at least in part, explained by the changing relative titer of juvenile and molting hormone. One might expect, for example, that the relative (and absolute) titer of these hormones is different at the RH-stage, which is at the end of a larval molt, from that at the later stages. Recent observations have given an important and unexpected clue that might provide an explanation for the tissue specificity of the hormonal response.

Among the first indications of hormone-induced metabolic changes in the gland cells is the formation of puff 18-C in the Ist and of puff 2-b in the IInd chromosome (Clever, 1961, 1962). We recently discovered that these two puffs differ in their sensitivity to the ecdysone analogues, α- and β-ecdysone: one of them, puff I-18-C, is sensitive only to α-ecdysone, the other one, puff IV-2-B, to β-ecdysone. The fact that both of them eventually appear after the injection of either analogue is due to the rapid conversion of each analogue to compounds with the activity of the other one (Clever et al., 1972). Rapid metabolism of injected ecdysones is a well established fact (cf., Robbins et al., 1971). Its physiological significance is little understood, but is usually discussed in connection with molting hormone catabolism. The endogeneous intracellular titers of the various analogues are not yet known. It is intriguing to speculate that the relative titers may differ between cell types and that the mechanism regulating the cell-type specific responses to the rising molting hormone titer may reside in the enzyme complement of each cell establishing its specific analogue spectrum.

ACKNOWLEDGEMENTS

The investigations reported here were supported by grants from the National Science Foundation. I gratefully acknowledge the collaboration of my colleagues, J. M. Darrow, P. A. Henrikson, L. Rubinstein, K. S. Schin and I. Storbeck, who made this work possible.

REFERENCES

Beermann, W. (1952). Chromosoma 5, 139.
Beermann, W. (1961). Chromosoma 12, 1.
Berger, E., Kafatos, F.C., Felsted, R.C. and Law, J.H. (1971). J. Biol. Chem. 246, 4131.
Bowers, W.E., Finkenstaed, J.T. and DeDuve, C. (1967). J. Cell Biol. 32, 339.
Clever, U. (1961). Chromosoma 12, 607.
Clever, U. (1962a). Chromosoma 13, 385.
Clever, U. (1962b). J. Insect Physiol. 8, 357.

Clever, U. (1963). Chromosoma 14, 651.

Clever, U. (1969). Exp. Cell Res. 55, 317.

Clever, U., Clever, I., Storbeck, I. and Young, N.L. (1972) (in preparation).

Clever, U. and Storbeck, I. (1970). Biochim. Biophys. Acta 217, 108.

Clever, U., Storbeck, I. and Romball, C.G. (1969). Exp. Cell Res. 55, 306.

Cohn, Z.A. and Fedorko, M.E. (1969). In "Lysosomes in Biology and Pathology". (J.T. Dingle and H.B. Fell, eds.), Vol. I, pp. 43-63. North Holland Publ. Co., Amsterdam.

Daneholt, B. and Edström, J.-E.(1967). Cytogenetics 6, 350.

Daneholt, B., Edström, J.-E., Egyhazi, E., Lambert, B. and Ringborg, U. (1969). Chromosoma 28, 399.

Daneholt, B. and Svedhem, L. (1971). Exp. Cell Res. 67, 263.

Darnell, J.E., Pagoulatos, G.N., Lindberg, U. and Balint, R. (1970). Cold Spring Harbor Symp. quant. Biol. 35, 555.

Darrow, J.M. and Clever, U. (1970). Develop. Biol. 21, 331.

Doyle, D. and Laufer, H. (1969a). J. Cell Biol. 40, 61.

Doyle, D. and Laufer, H. (1969b). Exp. Cell Res. 57, 205.

Ellgaard, E.G. and Clever, U. (1971). Chromosoma 36, 60.

Grossbach, U. (1969). Chromosoma 28, 136.

Harrod, M.J.E. and Kastritsis, D.C. (1972). J. Ultrastr. Res. 38, 482.

Henrikson, P.A. and Clever, U. (1972). J. Insect Physiol. (in press).

Keyl, H.G. and Pelling, C. (1963). Chromosoma 14, 347.

Kloetzel, J.H. and Laufer, H. (1969). J. Ultrastr. Res. 29, 15.

Laufer, H. (1968). Am. Zool. 8, 257.

Laufer, H. and Schin, K.S. (1971). Can.Ent. 103, 454.

Lockshin, R.A. (1969). J. Insect Physiol. 15, 1505.

Lockshin, R.A. and Williams, C.M. (1965a). _J. Insect Physiol._ 11, 601.

Lockshin, R.A. and Williams, C.M. (1965b). _J. Insect Physiol._ 11, 803.

Ohnischi, E., Dohke, K. and Ashida, M. (1970). _Archs. Biochem. Biophys._ 139, 143.

Plagemann, P.G.W. (1971). _J. Cell. Physiol._ 77, 241.

Radford, S.V. and Misch, D.W. (1971). _J. Cell Biol._ 49, 702.

Robbins, W.E., Kaplanis, J.N., Svoboda, J.A. and Thompson, M.J. (1971). _Ann. Rev. Entomol._ 16, 53.

Rodems, A.E., Henrikson, P.A. and Clever, U. (1969). _Experientia_ 25, 686.

Rodman, T.C. (1968). _Chromosoma_ 23, 271.

Rubinstein, L. and Clever, U. (1972). _Develop. Biol._ 27, 519.

Sachs, R.I. and Clever, U. (1972). _Exp. Cell Res._ (in press).

Saunders, J.W. (1966). _Science_ 154, 604.

Scherrer, K. and Marcaud, L. (1968). _J. Cell. Physiol._ 72, Suppl. 1, 181.

Scherrer, K., Spohr, G., Granbulan, N., Morel, C., Grosclaude, J. and Chezzi, C. (1970). _Cold Spring Harbor Symp. quant. Biol._ 35, 539.

Schin, K.S. and Clever, U. (1965). _Science_ 150, 1053.

Schin, K.S. and Clever, U. (1968a). _Z. Zellforsch._ 86, 262.

Schin, K.S. and Clever, U. (1968b). _Exp. Cell Res._ 49, 208.

TABLE 1

Effect of actinomycin D and
cycloheximide on replication

| Age treatment | % Nuclei labeled | | |
started	Actinomycin	Cycloheximide	Control
RH, $<$12 hours	0.9 ± 0.4	0.2 ± 0.1	12.6 ± 1.7
RH, total	1.0 ± 0.3	1.3 ± 0.4	12.6 ± 1.7
Day 3	12.3 ± 1.4	10.0 ± 1.0	11.1 ± 1.3
Day 5	-	7.5 ± 1.3	9.2 ± 1.2
Day 8	6.0 ± 1.1	4.1 ± 0.8	6.5 ± 0.8
Day 10	2.2 ± 0.6	2.0 ± 0.5	3.7 ± 0.9

Drug treatment was for 48 hours in all experiments,
after which glands were explanted and incubated with
^{3}H-thymidine (10 µCi/ml) for 30 minutes. Actinomycin
was used in a concentration of 2 µg/ml, cycloheximide
of 10 µg/ml. After Darrow and Clever (1970).

TABLE 2

Intracellular distribution of the pH 3.5 protease

Stage	Homogenization medium	Percentage P-10	Activity P-100	Recovered in Fraction S-100
prepupae	0.25 M sucrose	69	25	6
prepupae	Distilled water	18		82
pupae	0.25 M sucrose	71		29
Opp-3, CHM	0.25 M sucrose	>95		<5

Gland homogenates were subjected to differential centrifugation. Low speed supernatants were discarded. Pellets formed after centrifugation at 10,000 x g (P-10) and 100,000 x g (P-100) and the 100,000 x g supernatant (S-100) were assayed for enzyme activity. The last line refers to the experiments (described in the text) in which old prepupae were placed into cycloheximide for 24 hours before fractionation. After Henrikson and Clever (1972).

TABLE 3

Effect of trypsin on pH 3.5 protease activity in gland homogenates

Stage	Enzyme activity (OD/3hr/gland)		
	Control	Trypsin treated	% increase
larvae	0.052	0.053	2
larvae/E	0.049	0.064	31
larvae/EA	0.024	0.024	0
mpp	0.084	0.117	40
opp-1	0.102	0.156	53
pupae	0.180	0.198	10
mpp/P-10	0.174	0.260	49
mpp/S-100	<0.010	<0.012	-
opp/3/P-10	0.178	0.170	-5

Samples of gland homogenates (or fractions P-10 or S-100, see Table 2) were treated with trypsin (0.1 mg/ml; pH 7.5) for 10 minutes before assaying for protease activity at pH 3.5. Untreated samples of the same homogenates served as controls. Larvae in lines 2 and 3 were injected with 0.4 µg of β-ecdysone and kept in normal culture medium (E) or actinomycin solution (EA) for 48 hours. From Henrikson and Clever (1972).

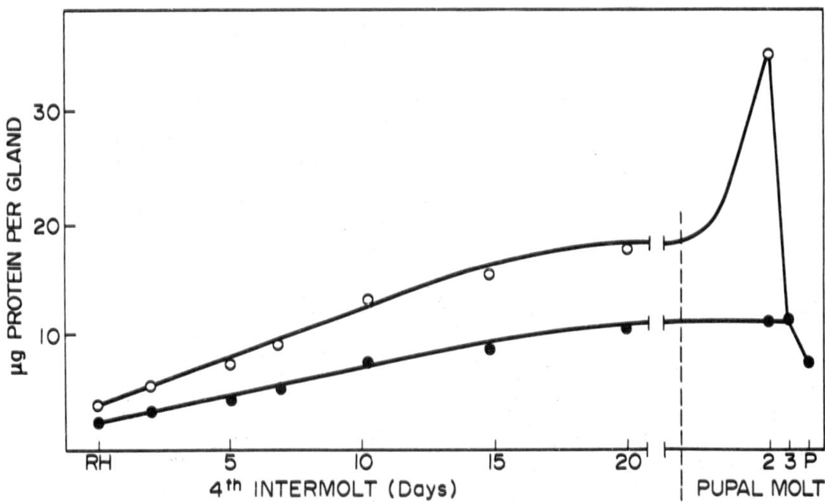

Fig. 1 Growth of the salivary gland of
Chironomus during the last larval instar. Open circles refer to total glandular protein, including the
secretory proteins of the glandular lumen; closed
circles refer to only cellular proteins. Where both
values are given the latter was estimated from the
former assuming that on the average 40% of the total
protein consists of secretory proteins (Darrow and
Clever, 1970). 2, 3 and P refer to molting stages
opp-2, opp-3 and young pupae as defined in the text.

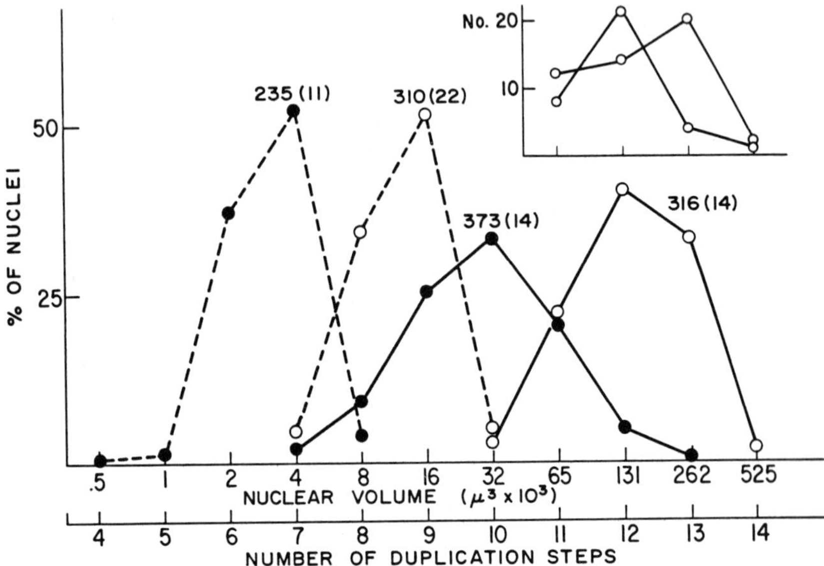

Fig. 2 Nuclear volumes in salivary gland (solid lines) and Malpighian tubule (broken lines) cells at the RH-stage (filled circles) and in 20-day-old last instar larvae (open circles). Volumes were grouped according to duplication steps of the 32 μ^3 volume of a diploid nucleus. The figures indicate the number of nuclei (and number of larvae) of each stage examined. The inset gives nuclear volumes in two individual glands. (From Darrow and Clever, 1970.)

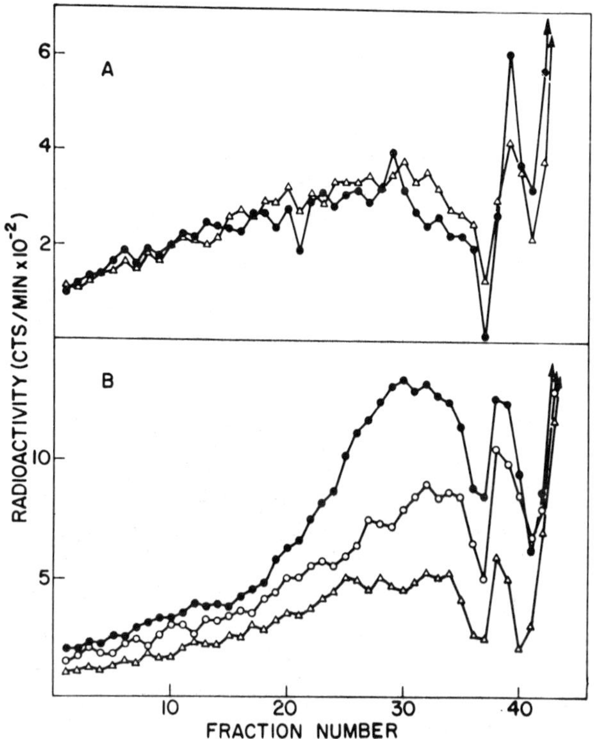

Fig. 3 Effect of actinomycin D on polyribosome profiles ($\triangle$) from glands of 10-day-old larvae (A) or red-head larvae (B). 0-hour (0-0) or timed controls (●-●) were used as described in the text. Experimental larvae were kept in actinomycin (4 µg/ml) for 24 hours before glands were explanted and incubated with ^{3}H-leucine. Extracts in A were prepared from 16 glands, those in B from 32 glands for each experimental group. (From Clever and Storbeck, 1970, which should be consulted for experimental details.)

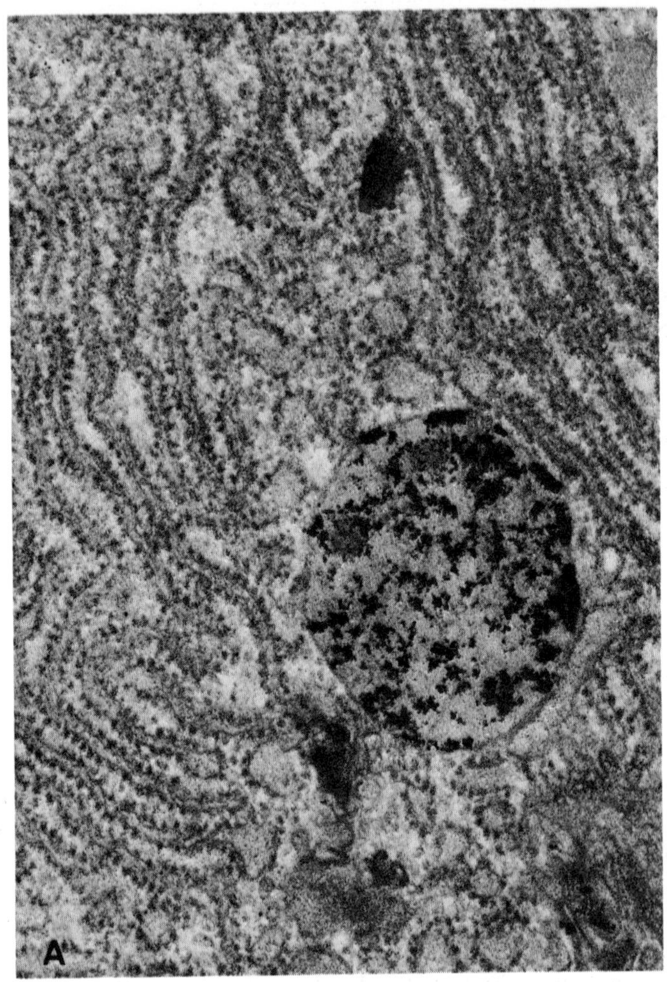

Fig. 4a   Distribution of acid phosphatase activity in a gland of a prepupa (A). In the prepupa,
the enzyme reaction products (lead deposits) are restricted to the lysosome. The glands were treated
with lead phosphate by a modified Gomori technique.
A, 82,000 x (From Schin and Clever, 1968).

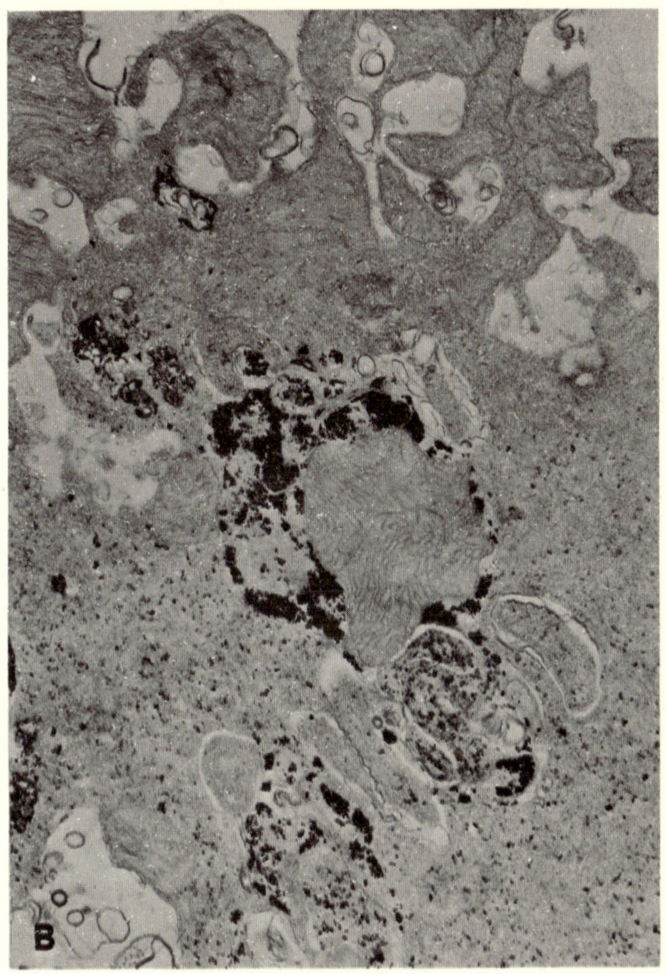

Fig. 4b Distribution of acid phosphatase activity in a gland of a young pupa (B). In the pupa,
the enzyme seems to be freely distributed in degen-
erated areas of the cell. Note also that in the
pupal gland the lysosomes appear larger and filled
with remnants of cellular structures such as mitochondria and myelin-like figures. The glands were
treated with lead phosphate by a modified Gomori
technique. B, 12,000 x (From Schin and Clever,1968).

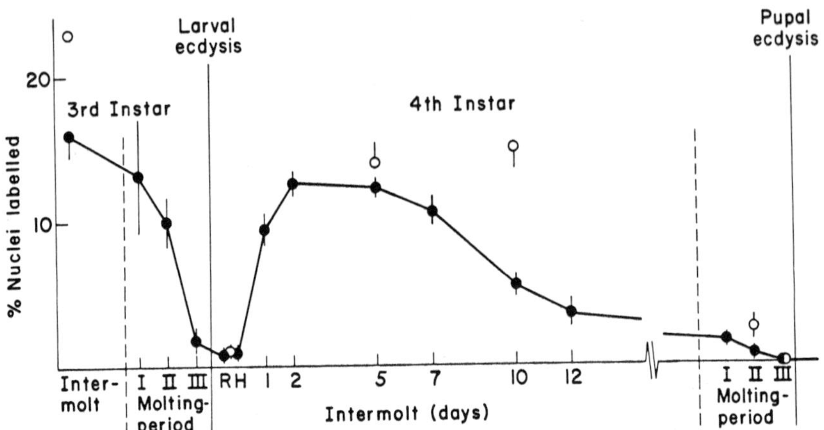

Fig. 5 Frequency of replicating nuclei in salivary glands at different stages of development. Explanted salivary glands were incubated with ^{3}H-thymidine and the percentage labeled nuclei was determined by autoradiography. Open circles refer to replication frequency in Molpighian tubules. (From Darrow and Clever, 1970.)

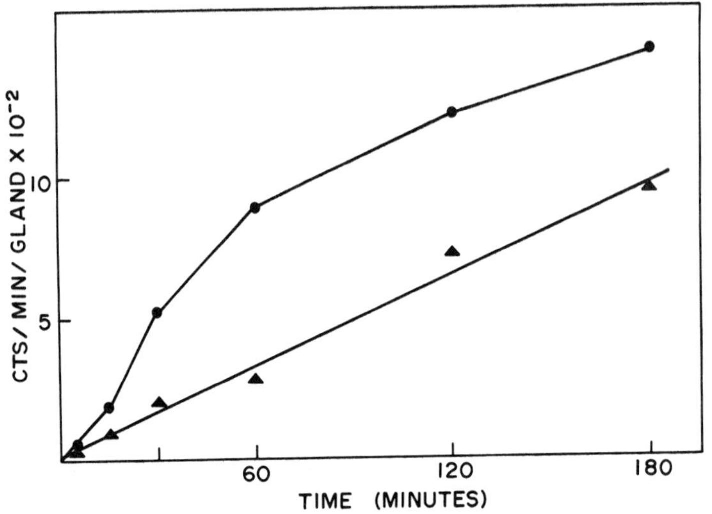

Fig. 6 Time course of ^{3}H-uridine incorporation into RNA of glands from 10-day-old larvae (●) and RH-larvae (▲). (From Rubinstein and Clever, 1972.)

66

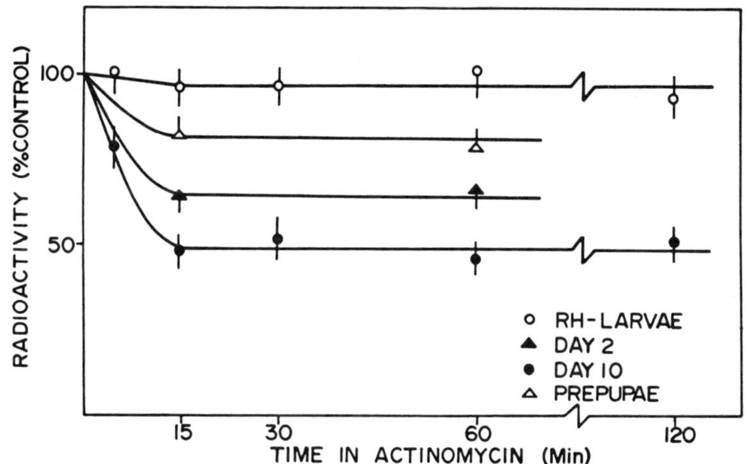

Fig. 7   Stability of newly synthesized RNA after inhibition of RNA synthesis with actinomycin D. Glands were incubated for 15 minutes with ^{3}H-uridine (100 µc/ml).   One gland of each pair served as control.   (From Rubinstein and Clever, 1972.)

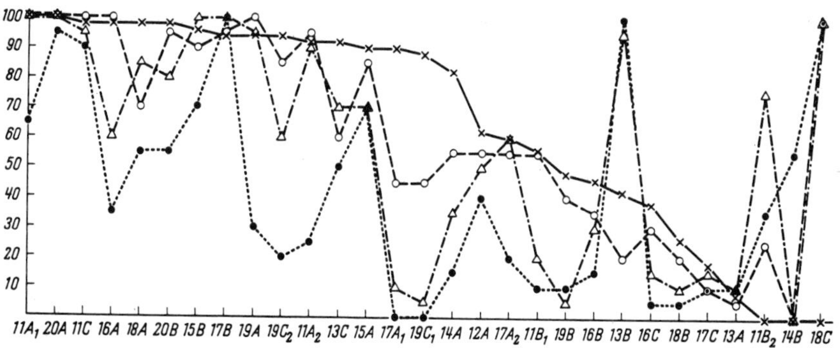

Fig. 8   Percentage larvae showing puffing (ordinate) at 29 loci of the Ist chromosome (absizissa) at different stages of the last larval instar. (x) intermolt larvae, (O) young prepupae, (△) mid-prepupae, (●) old prepupae.   20 larvae were examined for each stage.   (From Clever, 1962.)

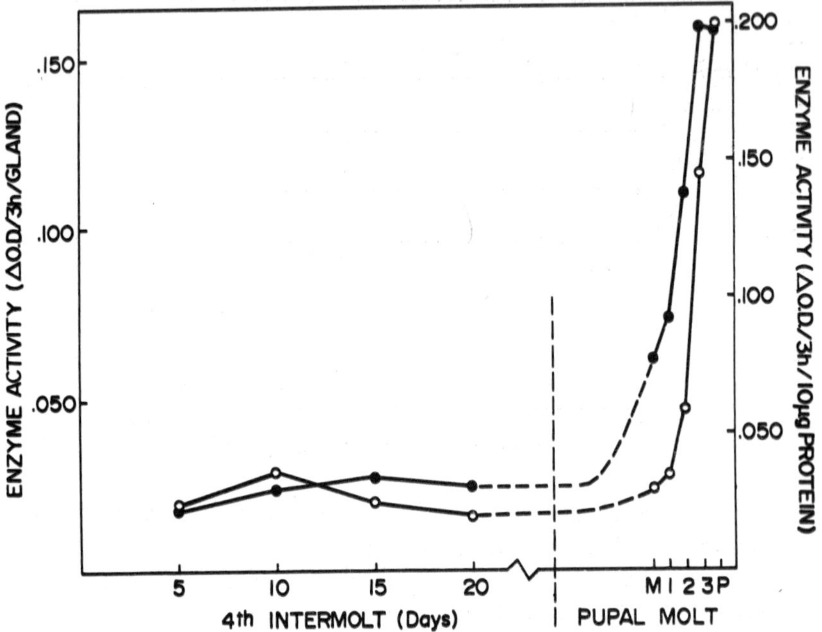

Fig. 9   pH 3.5 protease activity in glands from larvae at different stages of development. M, mid-prepupae; 1, 2, 3, old prepupae of stages opp-1, opp-2 and opp-3 (see text); P, young pupae. (●) enzyme activity per gland, (0) enzyme activity per 10 μg glandular protein (specific activity). (From Henrikson and Clever, 1972.)

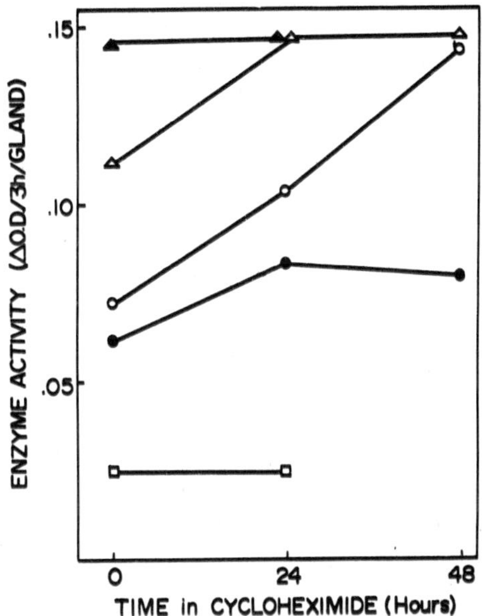

Fig. 10 pH 3.5 protease activity in glands from larvae kept in cycloheximide (10 µg/ml) for the periods of time indicated. (□) 10-day-old larvae, (●) mid-prepupae, (0) opp-1 prepupae, (△) opp-2 pre-pupae, (▲) young pupae. (From Henrikson and Clever, 1972.)

A DNA REPLICATION INTERMEDIATE
IN MOSQUITO DEVELOPMENT

Calvin A. Lang

Department of Biochemistry and
the Biological Aging Program
University of Louisville School of Medicine
Louisville, Kentucky

A. INTRODUCTION

The detailed knowledge of DNA replication is one
of the least well known areas of molecular biology.
This is especially true for eukaryotic organisms.
One of the most promising findings which suggested
that DNA replication is a discontinuous process was
the discovery by Okazaki of small DNA molecules in
bacterial systems (Okazaki et al., 1968).

Similar results have been found in a number of
mammalian systems including regenerating rat liver
(Tsukada et al., 1968), cultures of Chinese hamster
cells (Schandl and Taylor, 1969), Hela cells (Habener
et al., 1969), and human heteroploid cells (E.U.E.)
(Nuzzo et al., 1970). In none of these cases were
the DNA fragments isolated and characterized.

One of the most pertinent investigations to our
work was the finding of double-stranded 5-8 S DNA
fragments in the cytoplasm of cell cultures of embryonic mouse liver cells (Williamson, 1970). These
fragments were isolated and characterized and found
to have the same base composition as nuclear DNA.
The author concluded, however, that the DNA fragments
were breakdown products of nuclear DNA although he
mentioned the possibility of a precursor role. To
our knowledge this is the only work, other than our
own, in which DNA fragments from eukaryotic cells

have been isolated and characterized.

Our finding of a DNA replication intermediate in the developing mosquito larva evolved from earlier studies on various biochemical changes occurring during the life span of the laboratory mosquito. In this paper we are summarizing the investigations that led us to the discovery and characterization of this unusual DNA species which we term sDNA since it is found in the soluble fraction of tissue homogenates. Further we have obtained evidence from pulse-chase experiments that sDNA is a precursor and thus, a replication intermediate of nuclear DNA.

These results represent a number of different experimental approaches which involve investigation at different levels of biological organization. All of these approaches were important and are presented in the following narrative as the biology, biochem- istry, and molecular biology of the mosquito.

B. BIOLOGY OF THE LABORATORY MOSQUITO

The choice of the mosquito, _Aedes aegypti_, as a laboratory model for these kinds of studies has certain advantages which have been described earlier (Lang _et al_., 1965). Genetic control was achieved since inbred colonies were maintained for over 10 years representing at least 200 generations. Environ- mental control of temperature and lighting conditions, an important consideration for this poikilothermic organism, was readily accomplished with programmed incubators. A unique feature of environmental con- trol, namely, the nutritional history, was our devel- opment of a method to culture mosquito larvae under axenic (germfree) conditions on a medium that is almost completely defined chemically. Further, by restricting the adult diet to sucrose, ovarian devel- opment was prevented.

These features were incorporated into a stand- ardized procedure for the routine and also the axenic culture of mosquito larvae and subsequent develop- mental stages. Of particular significance has been

the technique of producing first instar larvae within
minutes and thus obtaining synchronized cultures
readily. This natural synchrony has enabled us to
discover in the metabolism of the whole organism
subtle changes which might otherwise be undetectable
in asynchronous cultures. The developmental times
of larvae and survival times of adults have been
determined with individuals as well as with populations of mosquitoes (Lang et al., 1972).

C. BIOCHEMICAL BACKGROUND

The biochemistry and intermediary metabolism of
the mosquito are similar to that of mammals. Physio-
logically they are aerobic organisms which reproduce
bisexually. Nutritionally they have dietary requirements for essential amino acids, vitamins, and
minerals. In addition, they require a dietary sterol
for development. The presence of a number of enzymes,
involving different metabolic pathways, further indicates the similarity of mosquitoes to mammals (Lang,
1967).

Our initial investigation with the mosquito was
concerned with biochemical parameters of cell number
during the developmental and aging periods which
comprise the entire life span of the mosquito (Lang
et al., 1965). To this end the concentrations of
total protein, RNA, and DNA were determined in the
whole organism, various body regions, and in subcellular fractions.

A primary consideration of these studies was
the validation of existing analytical procedures for
these macromolecular components. The Burton method
(1956) for the determination of DNA was modified by
the development of fractionation procedures to remove
substances which gave false positive results. These
contaminants were eliminated by proper extraction
procedures.

The results of this initial study indicated that
the profiles for weight, protein, RNA, and DNA were
superimposable except for the first two days of

larval life.

The most surprising finding was a high concentration of DNA in the high speed supernatant fraction of larval homogenates and not in adult homogenates. The percentage of total DNA found in that fraction was 31%. Further investigation of the percentage sDNA as a function of age demonstrated that this high concentration was correlated with the period of rapid larval growth and the percentage decreased to 2-3% in the late larval period and remained at the low level throughout the pupal and early adult stages (Lang and Meins, 1966).

Preliminary experiments carried out at that time indicated that sDNA was indeed a DNA moiety according to its susceptibility to DNase hydrolysis, acid insolubility, and its UV spectrum.

D. MOLECULAR BIOLOGY

The finding of sDNA was of interest to us for two reasons. First, its occurrence was correlated with growth. Second, its appearance in the soluble fraction suggested that it had unusual properties which could be different from those of nuclear DNA. Although it was not our initial objective, we felt that this finding deserved further inquiry.

The key question was "What is the role of sDNA?". We approached this problem in two ways: 1) the isolation and characterization of both sDNA and nuclear DNA, which we termed pDNA, and 2) the investigation of the metabolism of these two molecular species by radiolabeling experiments _in vivo_ (Kao, 1971).

Isolation and Characterization - The isolation of sDNA presented certain technical difficulties which required the development of a new procedure. Standard procedures for the isolation of bacterial DNA's and insect DNA's had been applied without success. An important feature of our procedure was the avoidance of repeated phenol extraction steps which resulted in losses and low yields. Purified sDNA was obtained as a flocculent precipitate which could not be spooled like other kinds of DNA preparations. This property suggested that sDNA was in a less polymerized state than pDNA.

Both sDNA and pDNA were characterized by several different criteria. The base composition was determined by acid hydrolysis and quantitative thin layer chromatography. This procedure also demonstrated that RNA was absent as evidenced by the absence of uracil. It should be noted also that the preparation of these DNA species included treatment with RNase and pronase. Another criterion used to characterize sDNA and pDNA was the differential spectrophotometric technique of Hirschman and Felsenfeld (1966), which is quite sensitive to the presence of RNA and protein contaminants. A third procedure was the determination of buoyant density. All three procedures gave the same base composition for both DNA's, namely, 39.5 ± 0.915 (5) mole % (G + C) for sDNA and 39.1 ± 0.515 (8) for pDNA (Mean $\pm$ SEM (n)).

Absorbance versus temperature curves of sDNA and pDNA were also determined. The T_m value for both DNA's was $84^\circ C$ which is equivalent to a 36 mole % (G + C) content. This base composition was somewhat lower than those obtained by other methods but is less meaningful, for it is an indirect determination. Approximately 45% hyperchromicity was observed for both DNA's.

It was necessary at this time to demonstrate that the labeled sDNA investigated in the metabolic experiments was the same species which was isolated and characterized. To this end a supernatant fraction from radiothymidine-labeled larvae was centrifuged to equilibrium in a CsCl density gradient. The resultant profile indicated that the acid-precipitated radioactivity was in a single peak. Further, this profile was superimposable on a profile of isolated sDNA analyzed in the same way. Another important aspect of this experiment was that it demonstrated that sDNA was not a satellite form nor was it mitochondrial or ribosomal DNA which have buoyant densities different from nuclear DNA. Sucrose

density gradient analysis indicated that the size of
sDNA was 7-8 S and of pDNA was > 20 S.

In summary, the results of the isolation and
characterization experiments demonstrated that sDNA
and pDNA have the same base composition and are
double stranded. They differ only in size.

The Metabolism of sDNA and pDNA - Our early
studies indicated that sDNA was not a degradation
product of pDNA as shown by the recovery of known
amounts of DNA added to different steps of the
processing and analytical procedures. However, this
possibility was still considered in light of the fact
that the two DNA's were related in composition. The
objective of this series of metabolic experiments
was to determine whether sDNA was a precursor, a
degradation product, or both, in its relationship to
pDNA.

Larvae were pulsed by administration of radio-
thymidine to the culture medium and chased in a non-
radioactive medium. The incorporation of radioactivity into sDNA was over 10 times faster than into
pDNA. This result provided additional evidence that
sDNA was not a degradation product of pDNA.

The results of the chase experiment indicated
that sDNA decreases concomitantly with an increase in
pDNA, suggesting that sDNA plays a precursor role.
These results were not definitive however, for they
did not demonstrate chain elongation, a requisite for
DNA replication.

Our next experiments, therefore, were sucrose
density gradient analyses of DNA from pulsed-chased
larvae. In these experiments larvae were homogenized
in phenol to inactivate nuclease destruction and
ensure complete recovery. Under these conditions
both sDNA and pDNA were followed together, but since
our previous experiments had demonstrated a 10 to 1
ratio of sDNA to pDNA, the results would indicate
mainly the fate of sDNA.

The results under neutral conditions indicated
an increase in S values after several hours of pulse.
A more definitive experiment was done in which the

preparation and the sucrose density gradient analyses were performed under alkaline conditions which would rupture hydrogen bonds and yield single-stranded fragments.

The results under alkaline conditions indicated that increasing amounts of 7 to 8 S DNA were synthesized up to 6 hours. After a one-day chase the radioactivity appeared in 12 S. After two days chase the size increased further to 19-22 S, which is the sedimentation coefficient of pDNA under alkaline conditions. These findings indicated that chain elongation was taking place and provided the evidence that sDNA is an intermediate in the replication of DNA.

E. SUMMARY

Our experimental evidence indicates that a low molecular weight DNA of approximately 500,000 Daltons is present during rapid growth of mosquito larvae. This DNA has been isolated and characterized and shown to have the same base composition as bulk DNA. It differs, however, in size. Metabolically it appears that sDNA is an intermediate in the replication of bulk DNA. Although sDNA appears in the supernatant fraction, our preliminary radioautographic evidence suggests that it is not cytoplasmic but rather is of nuclear origin. Our current efforts are aimed at elucidating this point as well as the enzymes involved. The significance of this DNA replication intermediate to cell turnover in development and aging remains to be determined.

ACKNOWLEDGEMENTS

The author is deeply indebted to Betty Jane Mills for her valuable assistance in preparing this manuscript and to P. C. Kao, Carl Beyer, and Fred Meins for their integral role in the investigative aspects. The support of the National Institutes of Health through research grants and a Career Development Award to C.A.L. is also gratefully acknowledged.

REFERENCES

Burton, K. (1956). Biochem. J. 62, 315.

Habener, J.F., Bynum, B.S., and Shack, J. (1969).
    Biochim. Biophys. Acta 195, 484.

Hirschman, S.Z. and Felsenfeld, G. (1966). J. Mol.
    Biol. 16, 347.

Kao, P.C. (1971). Ph.D. Dissertation, University of
    Louisville, Louisville, Kentucky.

Lang. C.A. (1967). J. Gerontol. 22, 53.

Lang, C.A., Basch, K.J., Storey, R.F. (1972). J. Nutr.
    (In Press).

Lang, C.A., Lau, H.Y., and Jefferson, D.J. (1965).
    Biochem. J. 95, 372.

Lang, C.A. and Meins, F. (1966). Proc. Nat. Acad.
    Sci., Wash. 55, 1525.

Nuzzo, F., Brega, A., and Falaschi, A. (1970). Proc.
    Nat. Acad. Sci., Wash. 65, 1017.

Okazaki, R., Ikazaki, T., Sakabe, K. Sugimoto, K.
    and Sugino, A. (1968). Proc. Nat. Acad. Sci.,
    Wash. 59, 598.

Schandl, E.K. and Taylor, J.H. (1969). Biochem.
    Biophys. Res. Comm. 34, 291.

Tsukada, K., Moriyama, T., Lynch, W.E., and Lieber-
    man, I. (1968). Nature 220, 162.

Williamson, R. (1970). J. Mol. Biol. 51, 157.

BIOCHEMICAL CHANGES IN NUCLEIC ACIDS AND PROTEINS
DURING OVARIAN MATURATION AND EGG DEVELOPMENT

Ruth R. Painter

Department of Environmental Toxicology
University of California
Davis, California 95616

INTRODUCTION

I am pleased to present at this time some of our
work on the biochemical changes in nucleic acids and
proteins during ovarian maturation and embryonic
development of the house fly, _Musca domestica,_ L.
During the last 10 years we have been studying the
biochemical effects of insect chemosterilants. These
compounds are anti-fertility agents which may be of
value in the control of noxious insects. When they
are used in conjunction with pheromones or mass
release of sterile insects, it may be possible to
interrupt reproduction of a target species without
harming desirable ones (Borkovec, 1966; Kilgore,
1967; LaBrecque and Smith, 1968). Some compounds
which have been used in the study of nucleic acid
and protein synthesis in other organisms are effective insect chemosterilants, including anti-metabo-
lites, e.g., 5-fluorouracil and 5-fluorourotic acid,
and aziridinyl compounds, e.g., tepa, thiotepa and
apholate. These are some of the compounds we have
used in our studies.

When we started this project I reviewed the
literature and was surprised to find how little was
known about the biochemistry of any insect during
the reproductive cycle. Actually, this was an excellent time to become involved in such a project
because recent advances in microtechniques and

instrumentation have made it possible to carry out much of the work I am reporting today. Our research group has been small. It has usually consisted of Dr. W. W. Kilgore and myself, one other technician or graduate student, and a laboratory helper. Dr. Ahmed Gadallah and Dr. Khalid Al-Adil, as graduate students and as post-doctoral research assistants have contributed to this project.

After reviewing our results, I have selected for discussion today some of our work on (1) RNA and DNA synthesis including RNA-DNA hybridization, (2) ribosomes and protein synthesis by the ribosomes in vitro, (3) ribosomal RNA and ribosomal protein, and (4) proteins and some specific enzymes. First we will look at these compounds during oogenesis and then during embryogenesis. To contrast what we have found in the normal ovary and egg, I will include some of the results of interruption of reproduction with aziridinyl compounds. We have also used virgin females and unfertilized eggs in some of our studies.

The rearing procedures, treatment of diets, and preparation of samples have been previously described (Painter and Kilgore, 1967a; Gadallah, et al., 1970a). The flies were chemosterilized by including the selected compound in a dry diet of sugar, powdered egg, and dried milk for 36 hr after emergence. The flies were then maintained on the normal diet of diluted canned milk. After trying many methods including electric shock treatment of gravid females, we found by serendipity that we could obtain unfertilized eggs by not separating the males and females for 24 hr after emergence. The females for some unexplained reason are then stimulated to oviposit to a limited extent, although they have not mated. At the time of adult emergence, the testes contain mature sperm, but the ovaries are not fully developed (French and Hoopingarner, 1965) and the females will not allow mating until at least 48 hr after emergence.

Another important change we made in the usual rearing procedures involved altering the normal circadian rhythm for feeding and egg deposition by controlling the light and dark cycles. This change enabled us to be certain of getting the eggs laid within a predetermined 1-2 hr period in sufficient quantity for our experiments. At 24°C, the eggs hatch in _ca_. 15-18 hr. At 37°C, which is considered optimum temperature, the eggs hatch in 8 hr. Our observation on hatching times and temperature agree well with those of Davidson (1944) as cited by West (1951). Our embryogenesis studies were made on eggs collected shortly after deposition and incubated at 37°C for the desired time periods. The life cycle of the house fly is shown in Fig. 1. As far as I know there is no satisfactory method of separating males and females until after adult emergence. Therefore, we have limited most of our work to adults and eggs.

General Physical Characteristics. Table 1 is a summary of some physical measurements we have made on our colony of house flies (Painter and Kilgore, 1967b). The importance of adequate nutrition is readily seen when working with flies. Although the obvious variation is in the size of the adult fly, it is the quantity and quality of the larval media that is important in determining adult size. Adult females also require an adequate source of protein for ovarian development and normal oviposition. There can be considerable variation in the size and weight of eggs also. The smallest that I have weighed was 38 µg and the largest 67 µg. Over a period of years they have consistently averaged 52-55 µg.

NUCLEIC ACID SYNTHESIS DURING
OOGENESIS AND EMBRYOGENESIS

RNA and DNA and Intermediates. Fig. 2 shows the levels of RNA and RNA intermediates in the ovaries for 20 days after emergence (Gadallah, _et al_., 1970a). These were determined on the acid-soluble and acid-insoluble extracts by the orcinol colori-

81

metric procedure (Dische, 1955; Kilgore and Painter, 1964). We have tried a number of methods and extraction procedures, and although there are difficulties with all of them, this method seems to be reasonably good for insect tissues. And the results obtained by this simple method have been substantiated by more sophisticated techniques, as I will show later.

Even though we need additional data after day 5, it is evident that in the ovaries, after the initial rapid synthesis through day 3, there is a cyclic decrease and increase in both RNA and RNA intermediates. RNA is at a maximum at the time of first egg laying about day 5. These observations are consistent with the fact that by day 3, the ovaries are filled with mature oocytes which at the time of deposition and throughout embryogenesis maintain a relatively high level of RNA (Fig. 3) (Painter and Kilgore, 1967a). Bier (1963a,b) in his excellent studies showed that in _Musca_ and _Calliphora_, RNA and some protein are directed from the nurse cells through the cytoplasm into the oocytes. We have not determined the site of synthesis of the RNA, but from the work of Bier with _Musca_ and reports on other species (Engelmann, 1970), the nurse cells would appear to be the site of synthesis in _Musca_. The relatively high level of RNA intermediates in eggs (about twice that present in the ovaries at the time of egg deposition) is significant probably for synthesis of RNA for ribosomes and proteins in later stages in embryogenesis.

Figure 4 shows the levels of DNA and DNA intermediates in the ovaries as determined on the hot and cold $HClO_4$ extracts by the diphenylamine colorimetric method (Burton, 1956; Gadallah, _et al._, 1970a). Initial DNA synthesis is even more rapid than that of RNA. It reaches a maximum on day 2 and then sharply decreases. The decrease begins prior to egg deposition and is consistent with the very small quantity of DNA found in newly deposited fly eggs (Fig. 5) (Painter and Kilgore, 1967a). Again there is a cyclic increase and decrease during each oviposition cycle.

We should note that the results shown in Fig. 4 are calculated as µg DNA/mg of fresh tissue. For the first 4 days after adult emergence, the ovaries increase very rapidly in weight from about 0.2 mg/pair to about 5.25 mg/pair (Mitlin and Cohen, 1961). When the DNA is recalculated on the basis of the total amount present in the ovaries, the picture becomes slightly different. At the time of emergence the ovaries contain less than 0.05 µg DNA. By day 1, the DNA content has increased only to about 0.24 µg, but by day 2 the total has jumped to 3.6 µg. The maximum of <u>ca</u>. 8 µg DNA/pair of ovaries is reached on day 3.

Similar cycling of RNA and DNA has been found in the ovaries of <u>Aedes aegypti</u> (Dr. C. Judson, University of California, Davis, Dept. of Entomology, private communication).

The 700-fold increase in DNA in the eggs after an initial lag period of 1-2 hours shown in Fig. 5 represents my first exciting and satisfying series of experiments in insect biochemistry. Actually it was the result of slow, hard work for months that taught us how to control egg deposition, hatching procedures, and handling of samples to get reproducible results. These results also paved the way for much of our later research including the DNA-RNA hybridization studies.

When our first studies on DNA in fly eggs were published (Kilgore and Painter, 1964), only a limited amount of information on DNA synthesis in insect embryos was available. Lu and Bodine (1953) had studied changes in the distribution of phosphorus in the developing grasshopper (<u>Melanoplus differentialis</u>) embryo but had not determined the level of DNA in the embryo prior to day 15 of the 70 day developmental period. In 1963, Devi <u>et al</u>. reported on nucleic acid metabolism in <u>Tribolium confusum</u> including very limited data on DNA synthesis during embryogenesis. However, in 1954 Hoff-Jørgenson had published an

important paper on frog and sea-urchin eggs and
chick embryos which showed that after fertilization,
there was in each case an initial lag period followed by very rapid synthesis of DNA.

The low level of DNA intermediates found in the
eggs throughout embryogenesis raises some still unanswered questions. Lu and Bodine (1953) suggested
that RNA and RNA pool material might supply materials
from which DNA is synthesized. Devi, et al. (1963)
reported a similar low level of "free DNA nucleotides" in Tribolium eggs but did not comment on it.

Another method we have used to follow the
synthesis of nucleic acids is "methylated albumen
kieselguhr" (MAK) column chromatography. The nucleic
acid extracts were prepared essentially by the method
of Tissieres (1959) and the MAK column chromatography
was carried out by modification of the method of
Mandell and Hershey (1960) (Gadallah, et al., 1970a).
Peak areas indicate the presence of tRNA, DNA and
rRNA respectively. When radioactive precursors are
orally administered to the adult flies, the synthesis
of the nucleic acid species can be followed by liquid
scintillation counting of the fractions. Intermediates are swamped off the column with other low-
weight compounds and therefore are not visible in the
profile.

The MAK chromatographic profile of nucleic acids
in the extract of the ovaries is shown in Figure 6
(Al-Adil, 1972). The ovaries of newly emerged females are proportionally higher in tRNA than ovaries
from 3 to 6 day old females as determined by measurement of the areas under the peaks. Although the
quantity of DNA in the ovaries is small at all times
in relation to the amount of RNA present, DNA
synthesis is readily observed when the adult flies
are fed thymidine-5-triphosphate-H^3 (TTP-H^3) and the
TTP-H^3 incorporated into the newly synthesized DNA
is monitored by liquid scintillation counting of the
eluted fractions. When C^{14} labeled precursors of RNA
are also administered, the area of overlap in the DNA
and rRNA fractions is clearly visible in the nucleic

acid profile (Fig. 6). The profile of the nucleic
acids present in normal and chemosterilized eggs
after 0, 3, and 6 hr of incubation at 37°C is shown
in Figure 7 (Gadallah, et al., 1970a). The great
increase in DNA during embryogenesis is evident.
Little or no synthesis of DNA occurs in eggs from
aziridinyl chemosterilized flies.

  RNA-DNA Hybridization. To study further the
nucleic acids during oogenesis and embryogenesis,
we conducted a series of experiments in which RNA
and DNA were isolated and the degree of hybridization
in vitro was measured (Gadallah, et al., 1971a,b).
The formation of hybrids between RNA and denatured
DNA in vitro has been widely used as a test for the
complementarity of the two nucleic acids involved.
The degree of hybridization of two homologous or
heterologous forms has also been employed as a
measure of the complementarity of the two forms at
different stages of development and under different
conditions. Denis (1966) employed DNA-RNA hybridization to study the nucleic acids in the developing
embryos of Xenopus laevis. Similar methods were
used by Whitely, McCarthy and Whitely (1966) to
determine the changing populations of messenger RNA
before and after fertilization of sea-urchin eggs.
We have used the nucleic acids from ovaries and eggs
of mated and virgin females, and the ovaries and eggs
from mated normal and chemosterilized females for
preparation of the homologous and heterologous
hybrids.

  The RNA was labeled by feeding the house flies
relatively large doses of uracil-C^{14} (100 µCi/gm of
dry diet). The RNA was isolated from the desired
tissue by modification of the sodium dodecyl sulfate-
cold phenol method of Brown and Littna (1964)
(Gadallah, et al., 1971a,b). Modification of the
method of Ritossa and Spiegelman (1965) was used for
the isolation of the DNA. Two methods of hybridization of the denatured DNA and RNA-C^{14} were employed. Hybridization in solution followed by MAK
column chromatography was conducted according to the

protocol of Mandell and Hershey (1960) with the modifications of Hayashi, et al. (1965). For hybridization with the DNA immobilized on membrane filters, the procedure of Gillespie and Spiegelman (1965), was used.

We appreciate the advice given to us by Professor Toshio Yamakawa, Department of Biochemistry, Tokyo College of Pharmacology in helping us with the hybridization studies.

Table 2 summarizes the hybridization of denatured DNA and RNA-C^{14} prepared from fertilized and unfertilized eggs with the DNA immobilized on membrane filters (Gadallah, et al., 1971a). The degrees of hybridization of the homologous forms from the two types of eggs 1-2 hr and 7-8 hr after egg deposition, are similar. The amounts of heterologous forms from 1-2 hr eggs (fertilized DNA-unfertilized RNA hybrid and vice versa) were slightly lower, but we can only speculate at this time as to whether the differences are significant. However, when the hybrids of fertilized eggs from early and late embryonic periods were prepared, there was a 20% reduction in hybridization, and an even greater reduction when early and late hybrids of fertilized and unfertilized eggs were prepared. We interpret these results as showing that with fertilization and embryogenesis there is synthesis of additional forms of RNA and DNA needed for development, although some of the original forms are still present. The decrease in hybridization of the heterologous forms from unfertilized eggs after 6 hrs of incubation probably indicates that some degradation of the nucleic acids has occurred.

When the formation of DNA-RNA hybrids from fertilized normal and chemosterilized eggs was studied (Table 3), we found an entirely different picture (Gadallah, et al., 1971b). The homologous hybrids with both DNA and RNA from either normal or chemosterilized eggs, whether freshly deposited or incubated for 6 hr at 37°C, were formed to about the same degree. Heterologous hybrid formations with

86

DNA from normal eggs and RNA from chemosterilized
eggs, or vice versa, was less than half that for the
homologous forms with DNA and RNA from the same types
of eggs. It was evident that the DNA and RNA from
the chemosterilized eggs were complementary to each
other but only partially complementary to the RNA
and DNA from normal eggs. Therefore, oral administration of an alkylating aziridinyl compound, i.e.,
thiotepa, to adult females altered both the RNA and
DNA synthesized during oogenesis. The results obtained by hybridization in solution were in agreement with those obtained by hybridization with immobilized DNA.

RIBOSOMES AND SYNTHESIS OF PROTEIN
BY RIBOSOMES <u>IN VITRO</u>

<u>Isolation and Identification of Ribosomes</u>.
Ribosomes from the ovaries and eggs were isolated by
means of differential centrifugation (Tissieres,
<u>et al</u>., 1959) and subjected to sucrose density
gradient analysis (SDG) (Britten and Roberts, 1960)
as described by Gadallah, <u>et al</u>. (1970b). A Spinco
Model E Ultracentrifuge with Schlieren and ultraviolet optics was employed for the analytical ultracentrifuge analysis (Gadallah, <u>et al</u>., 1971c). We
received helpful assistance from Dr. Nasr Marei,
Department of Environmental Toxicology, University
of California, Davis for this phase of the ribosomal
study. The rRNA of the ribosomes from the ovaries
can be labeled by feeding the adult flies orotic
acid-C^{14} and the ribosomal proteins by feeding l-
leucine-H^3. Figure 8 shows the distribution of
ribosomal forms after SDG analysis in the ovaries
of 4-day old females (Al-Adil, 1972). The profile
indicates that in the ovaries there is a preponder-
ance of polymeric forms containing at least 3 ribosomal units. Polysomes have been identified as the
forms most active in protein synthesis (Petermann,
1964).

The profile of the ribosomes from fertilized

87

and unfertilized eggs (Fig. 9) shows, on the other hand, that in the eggs the monomers predominate even in the late embryonic period (Gadallah, _et al._, 1971c). However, during embryonic development, the ratio of the polymers to monomers increased from 0.98:1.00 to 1.72:1.00. The Schlieren pattern of the ribosomal forms in fertilized eggs, which substantiated the SDG analysis, was used to calculate sedimentation coefficients. We further confirmed the identification of the ribosomal forms by electron microscopy. Although no comparable data on ribosomal aggregation during insect embryogenesis have been located, reports on sea-urchin eggs have shown that in unfertilized eggs, the ribosomes are primarily monomeric, but shortly after fertilization an increase in ribosomal aggregation can be detected (Monroy and Tyler, 1963; Candelas and Iverson, 1966).

In unfertilized eggs there was no evidence of ribosomal aggregation after egg deposition as based on the ratio of polymers to monomers. But in aziridinyl chemosterilized eggs, some ribosomal aggregation occurred as the ratio of polymers to monomers increased to 1.30:1.00 in eggs incubated for 6 hr (Gadallah, _et al._, 1971d). This fact is consistent with the reports that eggs of house flies chemosterilized with aziridinyl compounds undergo one or more meiotic divisions before embryonic death occurs (Fahmy and Fahmy, 1964; LaChance, _et al._, 1968; Matolin, 1969).

For the _in vitro_ protein synthesis studies, the ribosomes and pH 5 enzymes were prepared from the eggs as described by Campagnoni and Mahler (1967). One μCi 1-phenylalanine-C^{14} was added to the mixture of 19 amino acids to label the synthesized protein. Under our experimental conditions, 300 μg protein from the supernatant solution (pH 5 enzymes) and ribosomes equivalent to 300-400 μg ribosomal protein produced optimum results in the system.

The ability of the various polymeric forms from fertilized, unfertilized and aziridinyl-chemosterilized eggs to initiate protein synthesis _in_

vitro is shown in Fig. 10 (Gadallah, _et al._, 1971c,
d). The data show that for the 3 types of eggs, the
polysomes containing 3-5 ribosomal units were most
active in protein synthesis. The polysomes from
fertilized eggs had a much greater capacity to
initiate protein synthesis than either those from
unfertilized or chemosterilized eggs. Ribosomes
from chemosterilized eggs were 5X's more active
than those from unfertilized eggs.

Ribosomal RNA and Ribosomal Proteins. When
the rRNA, isolated from the ribosomes of ovaries
and eggs was subjected to SDG analysis, it was
calculated to consist primarily of 17.4S and 27.2S
particles (Al-Adil, 1972; Gadallah, _et al._, 1971e).
A very small peak which was calculated to contain
4.0S particles was present in the SDG profile of
rRNA from the ovaries. Whether this peak represents
an undegraded form of rRNA has been the subject of
considerable discussion. There was an increase in
rRNA during ovarium and egg development with the
greater increase being in the heavy particles (27.2S).

Insect rRNA is usually cited as being composed
of 18S and 28S particles although some rather wide
variations are found in the literature. Many of the
measurements were made on whole body samples.
Takahashi (1966) reported that the rRNA of the larval
fat body of _Philosomia cynthia_ consisted of 27S, 18S,
and 10S particles, and Ishikawa and Newburgh (1971)
found 28S and 18S components in the rRNA of the
posterior silk gland of the wax moth. In a recent
pertinent report, Frelinger and Roth (1971) reported
that after adult emergence, RNA synthesis in the
developing ovaries of _Culex fatigens_ was primarily
rRNA which was composed of 28S, 17.5S and 4-5S
particles.

Ribosomal protein also increased during embryogenesis but the acrylamide gel electrophoresis
pattern of the ribosomal proteins did not change
(Gadallah, _et al._, 1972b). This fact is in agreement
with the belief of many workers that there are no
qualitative differences in the ribosomal proteins

obtained from different parts or stages of development of the same organism (Yankofsky and Spiegelman,
1962; Aronson, 1963; Bielka, et al., 1967; Lyttleton,
1968). Because, after gel electrophoresis of
proteins, a single band may consist of 2 or more
proteins with the same electrophoretic mobility, the
results do not rule out the possibility that new
ribosomal proteins are synthesized during embryonic
development.

PROTEINS AND SOME SPECIFIC ENZYMES

Proteins. Amino acids and proteins and their
relationship to aging in insects will be discussed
by other speakers at this symposium. I am sure
their reports will help in the interpretation of our
results. In addition to studying the changes in
proteins in the ovaries and eggs, we have included
adult haemolymph in this phase of our work because
it has been established that insect haemolymph
supplies the developing oocytes with essential amino
acids and proteins (Laufer, 1960; Telfer, 1965;
Bodnaryk and Morrison, 1966, 1968). Soluble protein
was measured by the orcinol colorimetric procedure
(Lowry, et al., 1951). Acrylamide disc gel electrophoresis was carried out as described by Davis (1964)
with the modifications of Painter, et al. (1972).
The protein levels of the haemolymph of adult
males and females 0-21 days of age are given in
Table 4. The figures are the average of 2-5 determinations on composite haemolymph samples representing 2 to 10 individuals. I do not believe that
there is much value in these determinations except
as general indications of the range of protein in
the adult haemolymph, because in some later determinations on single individuals, I found values
ranging from 23 to 46 µg protein/µl haemolymph for
females all 21 days of age. The haemolymph of the
fly acts as a ready water reserve (Wigglesworth,
1965) and is influenced by the physiological state of
the individual (Florkin and Jenuiaux, 1964). There-

fore, the concentration of solids in the haemolymph can be expected to show wide variations. The soluble protein content of a haemolymph sample from a single individual, or a few individuals, is of little value without simultaneous determination of total solids in the haemolymph or total volume of the same individual. So far I have not been able to do this on house flies.

Gel electrophoresis protein patterns of the haemolymph of normal and chemosterilized females at 5, 10 and 15 days of age are shown in Fig. 11 (Gadallah, et al., 1972c). For the first 4 days after emergence during ovarian maturation, our results are similar to those reported in the detailed study of Bodnaryk and Morrison (1966) on the relationship between nutrition, haemolymph proteins, and ovarian development in Musca. After 5 days of age, the changes in the ovarian protein pattern are minimal in normal females, but when reproduction is interrupted by oral administration of aziridinyl compounds, the patterns become very different. The most obvious change is the gradual accumulation of proteins in 2 zones. The protein patterns of the ovaries from chemosterilized flies also vary greatly from the normal pattern with an accumulation of protein in one area (Fig. 12). This band area corresponds to the "sex-specific" protein identified by Bodnaryk and Morrison (1968). deLoof and deWilde (1970) have postulated hormonal control of the "vitellogenic female protein" in insects. In the light of these reports and our findings, we hope to determine if alkylating compounds, e.g., the aziridines, affect hormonal regulation of protein synthesis and metabolism in insects.

Enzymes. Among the enzyme systems we have surveyed during oogenesis and embryogenesis are lactic acid dehydrogenase, malic acid dehydrogenase, glucose-6-phosphate dehydrogenase, and esterases. We have studied lactic acid dehydrogenase (LDH) extensively because of the changes in LDH associated with ovarian and embryonic development and because

of the effect of aziridinyl chemosterilization on
the enzyme system (Kilgore and Painter, 1964;
Gadallah, et al., 1972a).

LDH activity increased very rapidly during
normal ovarian development from less than 0.01 μM
lactic acid/min/mg protein on day 1 to more than
1.00 μM/min/mg protein in fully developed ovaries
(Fig. 13). In the mature ovaries of chemosterilized
females, LDH activity was less than one-tenth of
that present in normal ovaries, although the gel
electrophoresis pattern of the LDH showed the same
change from 2 to 3 isozymes found in normal ovaries.
In normal eggs, LDH activity appeared to parallel
the synthesis of DNA, and when there was no synthesis
of DNA in chemosterilized eggs, there was no apparent
increase in LDH activity (Fig. 14). During normal
embryogenesis, the LDH isozymes increased from 2 to
5, but no increase in isozymes was found in the
chemosterilized eggs.

During the course of our work with LDH, I
identified the isozyme, LDH-X in micro-electro-
phoresis patterns of the testes of house flies
(unpublished data, this laboratory). This isozyme
has been found in the testes and sperm of a number
of other species (Zinkham, et al., 1964; Goldberg,
1963; Goldberg and Lerum, 1972). We have not found
any reports of LDH-X in other insects.

RESPIRATION RATE OF HOUSE FLY EGGS

Although not an integral part of the nucleic
acid and protein study, the respiration rates of the
3 types of eggs (normal, unfertilized and chemosterilized) help in interpretation of some of the results.
Fig. 15 shows the respiration rate of normal and unfertilized eggs after deposition (unpublished data,
this laboratory). The respiration rate of chemosterilized eggs is also measurable. It usually rises
more rapidly than that of unfertilized eggs, but also
decreases faster. Therefore, death does not occur
in either unfertilized or chemosterilized eggs until

after egg deposition.

SUMMARY AND CONCLUSIONS

In the ovaries of house flies during each oviposition cycle, there is an increase followed by a decrease of RNA, DNA and protein. Protein and RNA are maximal at the time of oviposition, but DNA decreases prior to egg deposition. Mating stimulates rRNA synthesis and ovarian maturation but does not result in the synthesis of new forms of RNA and DNA as measured by RNA-DNA hybridization _in vitro_ (unpublished data, this laboratory).

During embryogenesis, there is very rapid synthesis of DNA and LDH after an initial lag period of 1-2 hr. LDH isozymes increase from 2 to 5 during embryogenesis. RNA-DNA hybridization studies _in vitro_ indicate that fertilization results in synthesis of new forms of RNA and DNA in the eggs, but that some of the original forms are still present in the late embryonic period. The polymeric forms of the ribosomes from eggs and ovaries have been identified and their ability to initiate protein synthesis _in vitro_ has been confirmed. In each case, polysomes containing 3 or more ribosomal units are most active in protein synthesis.

Unfertilized eggs are often considered to be in an inactive or quiescent state. However, we have shown that there is limited synthesis of protein and of both RNA and DNA, probably for cell maintenance in unfertilized house fly eggs after deposition. They also have an appreciable O_2 respiration rate for about 4 hr after egg deposition. These results are in agreement with recent studies on unfertilized sea-urchin eggs, that contrary to previously accepted theories, some protein and nucleic acid synthesis continues in unfertilized eggs (Epel, _et al._, 1969; MacKintosh and Bell, 1969).

Permanent interruption of reproduction in the house fly by oral ingestion of aziridinyl compounds for 48 hr after emergence produces long-lasting

changes in the gel electrophoresis patterns of the
haemolymph and ovaries. In the ovaries, although
cyclic variations in RNA and DNA are not appreciably
affected, the pool of DNA intermediates is very low
and LDH activity is only one-tenth that of normal
ovaries. The eggs from chemosterilized flies have a
small, but measurable increase in RNA after deposition. DNA synthesis and LDH activity is almost
completely inhibited and there is an accumulation of
DNA intermediates. RNA-DNA hybridization studies
in vitro indicate that the DNA and RNA in chemosterilized eggs are complementary to each other, but
only to a limited degree to the RNA and DNA from
viable eggs.

ACKNOWLEDGEMENT

The author wishes to thank Dr. R. I. Krieger,
Dr. C. L. Judson, and Dr. K. M. Al-Adil for review-
ing the manuscript and Elaine Tillman for technical
assistance.
This work was partially supported by Research
Grant HD-01265 from the National Institutes of Health.

REFERENCES

Al-Adil, K.M. (1972). Ph.D. Thesis. University of
    Calif., Davis, Calif.
Aronson, A.I. (1963). Biochim. Biophys. Acta. 72,
    176.
Bielka, H., Welfe, H., and Schneiders, I. (1967).
    Acta Biol. Med. Ger. 19, K13-K14.
Bier, K. (1963a). J. Cell. Biol. 16, 436.
Bier, K. (1963b). Arch. Entwicklungsmech. Organismen.
    154, 552.
Bodnaryk, R.P., and Morrison, P.E. (1966). J. Insect
    Physiol. 12, 963.
Bodnaryk, R.P., and Morrison, P.E. (1968). J. Insect
    Physiol. 14, 1141.
Borkovec, A.B. (1966). "Insect Chemosterilants."
    Advan. Pest. Control Res. 7, Interscience, New

York.

Britten, R.J., and Roberts, R.B. (1960). _Science_ _131_, 32.

Brown, D.D., and Littna, E. (1964). _J. Mol. Biol._ _8_, 669.

Burton, K. (1956). _Biochem. J. 62_, 315.

Campagnoni, T.A., and Mahler, H.R. (1967). _Biochemistry 6_, 956.

Candelas, C.E., and Iverson, M.R. (1966). _Biochem. Biophys. Commun. 24_, 867.

Davidson, J. (1944). _J. Animal Ecol. 13_, 26.

Davis, B.J. (1964). _Ann. N.Y. Acad. Sci._ 121, 402.

deLoof, A., and deWilde, J. (1970). _J. Insect Physiol. 16_, 157.

Denis, H. (1966). _J. Mol. Biol._ 22, 269.

Devi, A., Lemonde, A., Srivastava, U., and Sarkar, N.K. (1963). _Exp. Cell. Res. 29_, 443.

Dische, Z. (1955). _In_ "The Nucleic Acids" (E. Chargaff and J.N. Davidson, eds.), Vol. I, pp. 285-305. Academic Press, New York.

Engelmann, F. (1970). "Physiology of Insect Reproduction." Pergamon Press, New York.

Epel, D., Pressman, B.C., Elsaesser, S., and Weaver, A.M. (1969). _In_ "The Cell Cycle: Gene-Enzyme Interactions" (G.M. Padilla, G.L. Whiteson, and I.L. Cameron, eds.), pp. 279-298. Academic Press, New York.

Fahmy, O.G., and Fahmy, M.J. (1964). _Trans. Roy. Soc. Trop. Med. Hyg. 58_, 318.

Florkin, M. and Jeuniaux, Ch. (1964). _In_ "The Physiology of Insecta" (M. Rockstein, ed.), Vol. III, pp. 109-152. Academic Press, New York.

Frelinger, J.A., and Roth, T.F. (1971). _J. Insect Physiol. 17_, 1401.

French, A., and Hoopingarner, R. (1965). _Ann. Entomol. Soc. Amer. 58_, 650.

Gadallah, A.I., Kilgore, W.W., and Painter, R.R. (1970a). _J. Econ. Entomol. 63_, 1777.

Gadallah, A.I., Kilgore, W.W., and Painter, R.R. (1970b). _J. Insect Physiol. 16_, 1245.

Gadallah, A.I., Kilgore, W.W., and Painter, R.R.

(1971a). _Insect Biochem._ 1, 302.

Gadallah, A.I., Kilgore, W.W., and Painter, R.R. (1971b). _Pestic. Biochem. Physiol._ 1, 166.

Gadallah, A.I., Kilgore, W.W., and Painter, R.R. (1971c). _Insect Biochem._ 1, 385.

Gadallah, A.I., Kilgore, W.W., and Painter, R.R. (1971d). _J. Econ. Entomol._ 64, 819.

Gadallah, A.I., Kilgore, W.W., and Painter, R.R. (1971e). _J. Econ. Entomol._ 64, 371.

Gadallah, A.I., Kilgore, W.W., and Painter, R.R. (1972a). _J. Econ. Entomol._ 65, 36.

Gadallah, A.I., Kilgore, W.W., Painter, R.R., and Marei, N. (1972b). _J. Econ. Entomol._ 65, 298.

Gadallah, A.I., Kilgore, W.W., and Painter, R.R. (1972c). _J. Econ. Entomol._ (In Press).

Gillespie, D., and Spiegelman, S. (1965). _J. Mol. Biol._ 12, 829.

Goldberg, E. (1963). _Science_ 139, 602.

Goldberg, E., and Lerum, J. (1972). _Science_ 176, 686.

Hayashi, M.N., Hayashi, M., and Spiegelman, S. (1965). _Biophys. J._ 5, 231.

Hoff-Jørgenson, E. (1954). _In_ "Recent Developments in Cell Physiology" (J.A. Kitching, ed.), Proc. 7th Symp. Colston Research Society, pp. 79-90, Academic Press, New York.

Ishikawa, H., and Newburgh, R.W. (1971). _Biochim. Biophys. Acta._ 232, 661.

Kilgore, W.W. (1967). _In_ "Pest Control, Biological, Physical and Selected Chemical Methods" (W.W. Kilgore and R.L. Doutt, eds.), pp. 197-239, Academic Press, New York.

Kilgore, W.W., and Painter, R.R. (1964). _Biochem. J._ 92, 353.

LaBrecque, G.C., and Smith, C.N. (1968). "Principles of Insect Chemosterilization." Appleton-Century-Crofts, New York.

LaChance, L.E., North, D.T., and Klassen, W. (1968). _In_ "Principles of Insect Chemosterilization" (G.C. LaBrecque and C.N. Smith, eds.), pp. 99-157. Academic Press, New York.

Laufer, H. (1960). _Ann. N.Y. Acad. Sci._ 89, 490.

Lowry, O.H., Rosebrough, N.J., Farr, A.L., and
    Randall, R.S. (1951). J. Biol. Chem. 193, 265.
Lu, L.K. and Bodine, J.H. (1953). Physiol. Zool. 26,
    242.
Lyttleton, J.W. (1968). Biochim. Biophys. Acta. 154,
    145.
MacKintosh, F.R. and Bell, E. (1969). Science 164,
    961.
Mandell, J.D., and Hershey, A.D. (1960). Anal.
    Biochem. 1, 66.
Matolin, S.(1969).Acta. Entomol. Bohemoslov. 66, 65.
Mitlin, N., and Cohen, C.F. (1961). J. Econ. Entomol.
    54, 651.
Monroy, A., and Tyler, A. (1963). Arch. Biochem.
    Biophys. 103, 431.
Painter, R.R., and Kilgore, W.W. (1967a). J. Insect
    Physiol. 13, 1105.
Painter, R.R., and Kilgore, W.W. (1967b). Ann.
    Entomol. Soc. Amer. 60, 1163.
Painter, R.R., Kilgore, W.W., and Gadallah, A.I.
    (1972). J. Econ. Entomol. 65, 23.
Petermann, M.L. (1964). "The Physical and Chemical
    Properties of Ribosomes." Elvisier Publishing Co.,
    New York.
Ritossa, F.M., and Spiegelman, S. (1965). Proc. Nat.
    Acad. Sci. U.S.A. 53, 737.
Takahashi, S. (1966). J. Insect Physiol. 12, 789.
Telfer, W.H. (1965) Ann. Rev. Entomol. 10, 161.
Tissieres, A., Watson, J.D., Schlessinger, D., and
    Hollingsworth, B.R. (1959). J. Mol. Biol. 1, 221.
West, L.S. (1951). "The Housefly." Comstock Publish-
    ing Co., Ithaca, New York.
Whitely, A.H., McCarthy, B.J., and Whitely, H.R.
    (1966). Proc. Nat. Acad. Sci. U.S.A. 55, 519.
Wigglesworth, V.B. (1965). "Principles of Insect
    Physiology" 6th Ed., p. 382, E.P. Dutton & Co.,
    New York.
Yankofsky, S.A., and Spiegelman, S. (1962). Proc.
    Nat. Acad. Sci. U.S.A. 48, 1069.
Zinkham, W.H., Blanco, A., and Clowry, L.J., Jr.
    (1964). Ann. N.Y. Acad. Sci. 121, 571.

TABLE 1

Average weights of egg, larva, chorion, pupa and adult house flies

	No. of Samples	Size of each Sample	Fresh Average weight (μg)	Dried Average weight (μg)	Percent of fresh wt.
			μg	μg	
Egg	20	100–200 eggs	55	14	25
Larva (newly emerged)	1	100 larvae	47	10	21
Chorion	1	100 chorions	5	2	40
Pupa			mg		
(normal)	100	1	18–21	--	--
(under-nourished	100	1	15–17	--	--
Adult					
Males (newly emerged)	100	1	10–14	--	--
Females (newly emerged)	100	1	13–18	--	--

TABLE 2

Hybridization of DNA-RNA from fertilized and
unfertilized house fly eggs

Type of nucleic acid	Degree of hybridization
a. Early developmental period hybrids	
f_0 DNA - f_0 RNA	0.260
μ_0 DNA - μ_0 RNA	.253
f_0 DNA - μ_0 RNA	.245
μ_0 DNA - f_0 RNA	.237
b. Late developmental period hybrids	
f_6 DNA - f_6 RNA	0.255
μ_6 DNA - μ_6 RNA	.247
f_6 DNA - μ_6 RNA	.193
μ_6 DNA - f_6 RNA	.187
c. Hybrids from early and late developmental periods	
f_0 DNA - f_6 RNA	0.205
f_6 DNA - f_0 RNA	.215
f_0 DNA - μ_6 RNA	.177
f_6 DNA - μ_0 RNA	.182
μ_0 DNA - μ_6 RNA	.239
μ_6 DNA - μ_0 RNA	.244
μ_0 DNA - f_6 RNA	.184
μ_6 DNA - f_0 RNA	.191

f_0 - from fertilized eggs, no incubation at 37°C
    (1-2 hr after oviposition)

μ_0 - from unfertilized eggs, no incubation at 37°C
    (1-2 hr after oviposition)

f_6 - from fertilized eggs, incubated 6 hr at 37°C
    (7-8 hr after oviposition)

μ_6 - from unfertilized eggs, incubated 6 hr at 37°C
    (7-8 hr after oviposition)

TABLE 3

Hybridization of DNA-RNA from the eggs of normal and thiotepa chemosterilized house flies[1]

Type of nucleic acid	Degree of DNA hybridization
n_0DNA-n_0RNA	0.248
c_0DNA-c_0RNA	0.237
n_0DNA-c_0RNA	0.095
c_0DNA-n_0RNA	0.098
n_6DNA-n_6RNA	0.236
c_6DNA-c_6RNA	0.228
n_6DNA-c_6RNA	0.091
c_6DNA-n_6RNA	0.095

[1] n_0 - from normal fertilized eggs, no incubation at 37°C (1-2 hr after oviposition);

c_0 - from thiotepa chemosterilized eggs, no incubation at 37°C (1-2 hr after oviposition);

n_6 - from normal fertilized eggs, incubated at 6 hr at 37°C (7-8 hr after oviposition);

c_6 - from thiotepa chemosterilized eggs incubated 6 hr at 37°C (7-8 hr after oviposition).

TABLE 4

Protein* content of the haemolymph of adult house flies

Age	Females[1]	Males[2]
	μg protein/μl haemoglobin	
4 hr	20	26
1 day	21	30
	29	32
	26	33
	26	25
3 days	54	43
	49	29
5 days	45	53
	42	34
8 days	32	--
	41	--
3 weeks	44	24
	52	32
	--	23

*Soluble Protein (Folin-Wu-Ciocalteau Method)
[1]2-5 Females
[2]2-10 Males

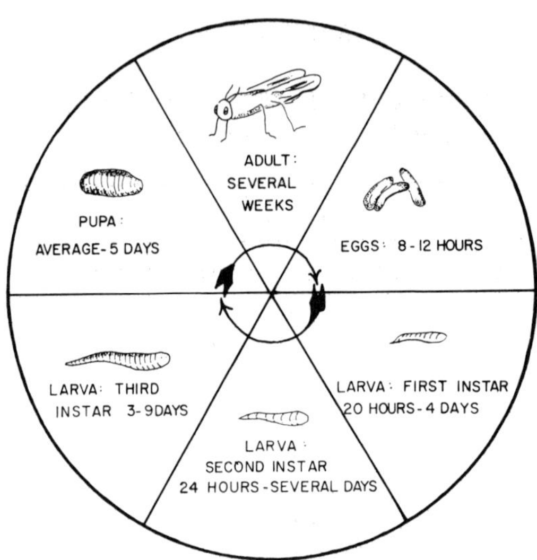

Fig. 1 Life cycle of the housefly (West, 1951. Reprinted by permission of the copyright owner).

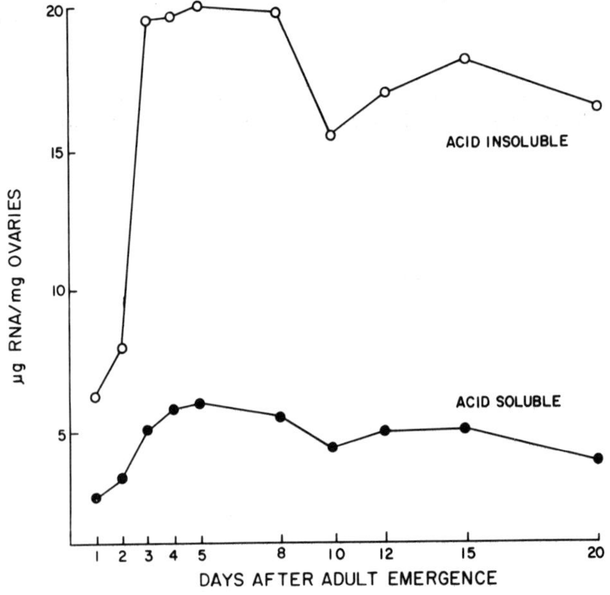

Fig. 2 RNA (acid-insoluble) and RNA intermediates (acid-soluble) of house fly ovaries.

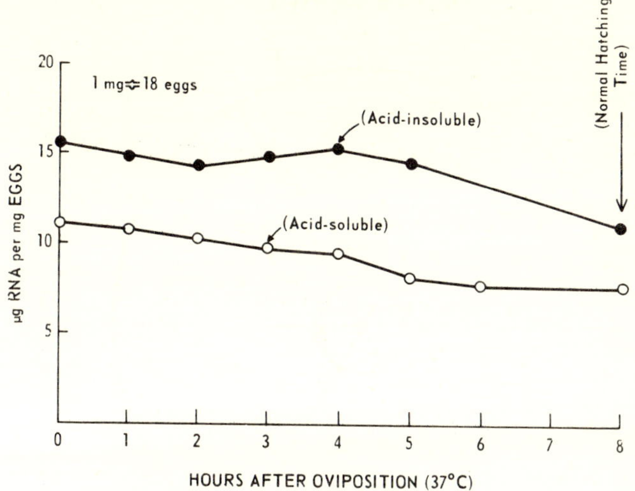

Fig. 3   RNA (acid-insoluble) and RNA intermediates (acid soluble) of house fly eggs (Painter
and Kilgore, 1967a.   Reprinted by permission of the
copyright owner).

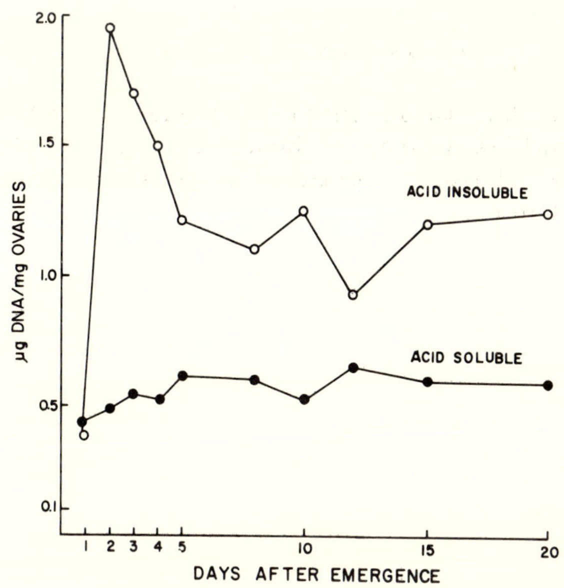

Fig. 4   DNA (acid-insoluble) and DNA intermediates (acid-soluble) of house fly ovaries.

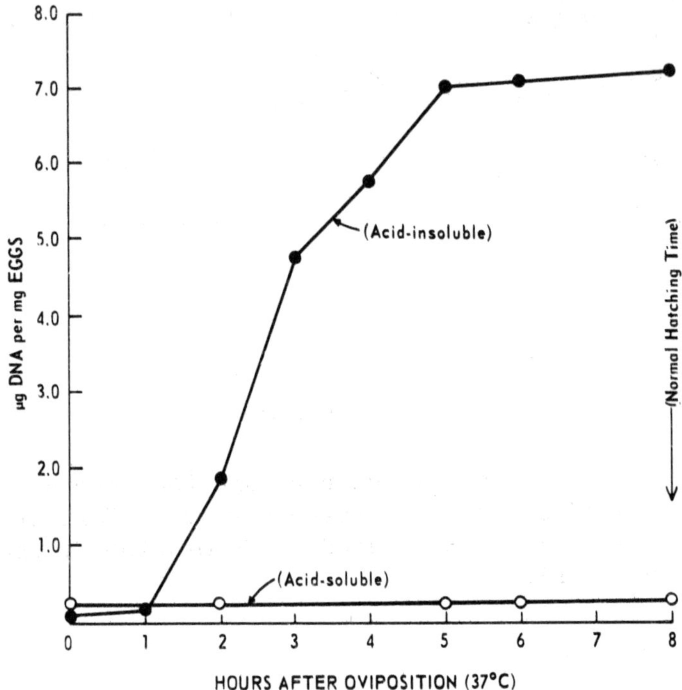

Fig. 5   DNA (acid-insoluble) and DNA intermediates (acid-soluble) of house fly eggs (Painter
and Kilgore, 1967a.   Reprinted by permission of the
copyright owner).

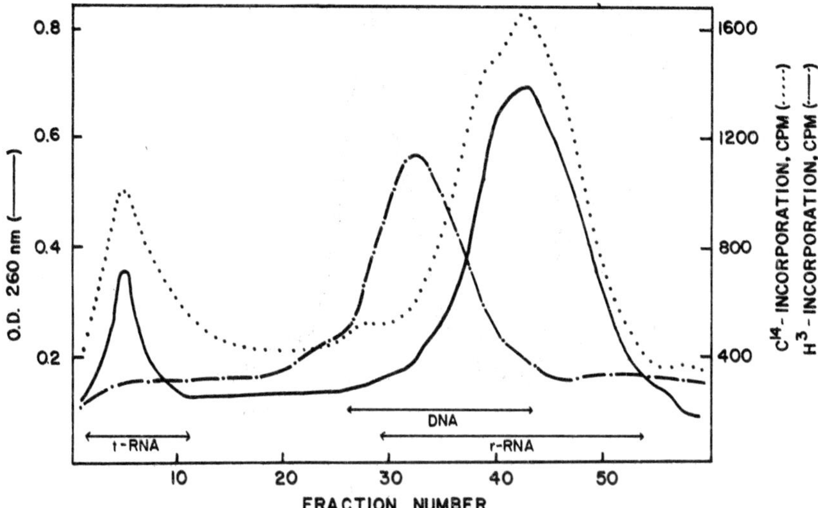

Fig. 6 Methylated albumen kieselguhr chromatographic profile of nucleic acids from the ovaries of house flies fed thymidine-5-triphosphate-H^3 and orotic acid-C^{14} to label DNA and RNA respectively.

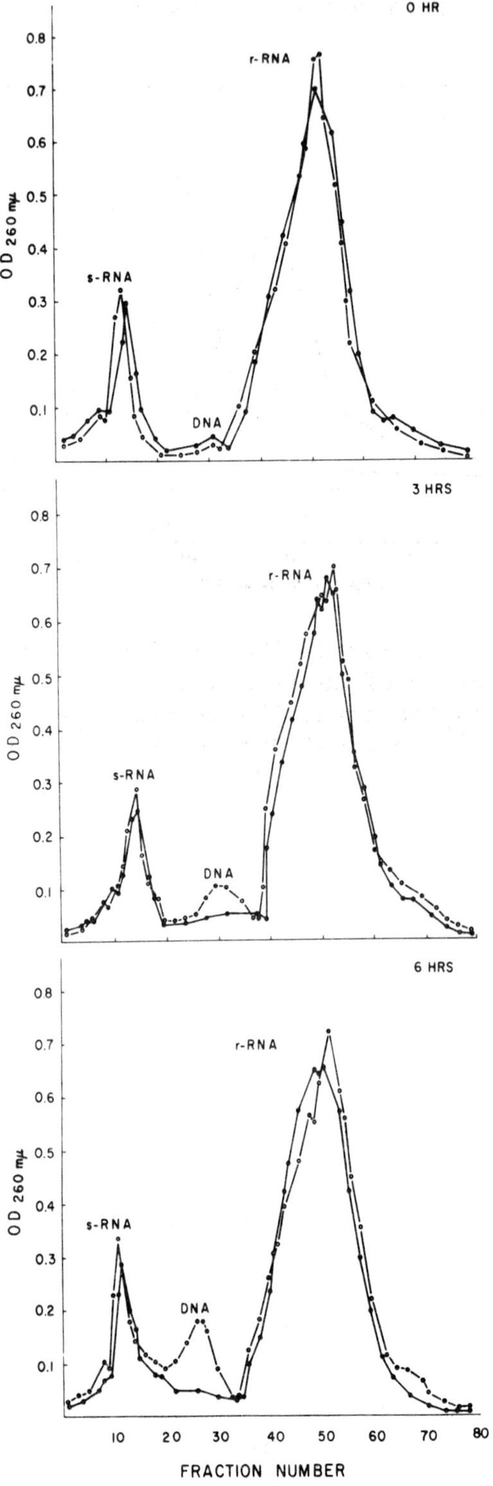

0 HR
OD 260 mμ
0.8
0.7
0.6
0.5
0.4
0.3
0.2
0.1
s-RNA
r-RNA
DNA
3 HRS
OD 260 mμ
0.8
0.7
0.6
0.5
0.4
0.3
0.2
0.1
s-RNA
DNA
r-RNA
6 HRS
OD 260 mμ
0.8
0.7
0.6
0.5
0.4
0.3
0.2
0.1
s-RNA
DNA
r-RNA
FRACTION NUMBER
10
20
30
40
50
60
70
80

Fig. 7 Methylated albumen kieselguhr chromatographic profiles of nucleic acids from the eggs of
house flies. o——o from normal house flies. ●——●
from chemosterilized house flies (Gadallah, _et al._,
1970a. Reprinted by permission of the copyright
owner).

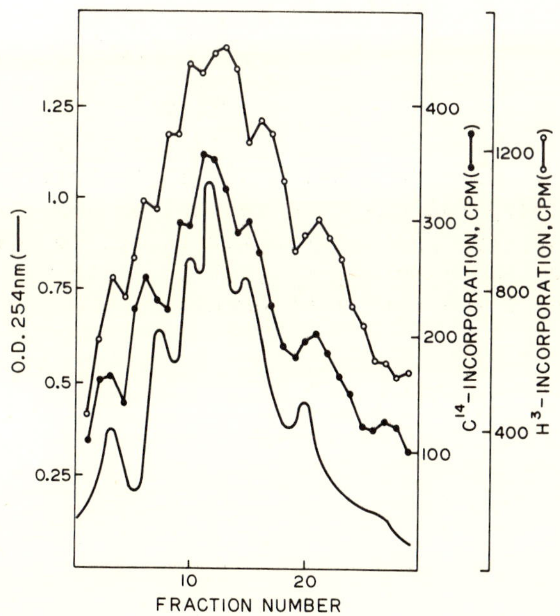

Fig. 8 Sucrose density gradient profile of a
ribosomal preparation from house fly ovaries. The
ribosomes were labeled by feeding the house flies
orotic acid C-^{14} and 1-leucine-H^{3}. From left to
right the peaks represent subunits, monomers, dimers,
trimers and heavier polymers.

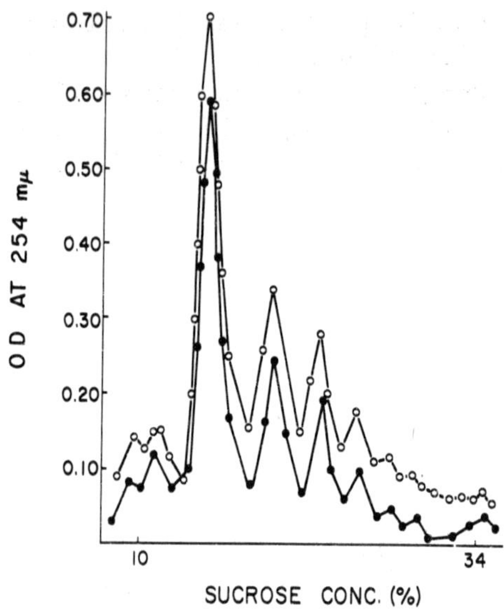

Fig. 9 Sucrose density gradient profile of ribosomal preparations from fertilized and unfertilized house fly eggs. o——o fertilized; ●——● unfertilized (Gadallah, et al., 1971c. Reprinted by permission of the copyright owner).

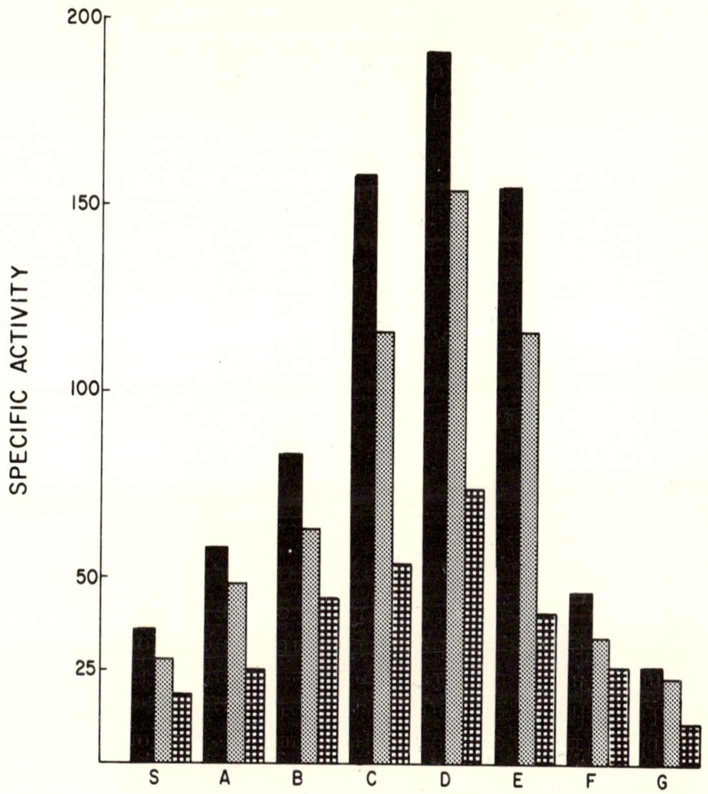

Fig. 10   Protein synthesis in a cell-free system, utilizing ribosomal units prepared from fertilized, unfertilized and chemosterilized eggs measured by incorporation of 1-phenylalanine-C^{14}. Ten drops from each well-defined peak area after SDG separation were used as the source of ribosomes. Specific activity in μμmoles 1-phenylalanine per mg protein.   S, subunits; A, monomers; B, dimers; C, trimers; D, E, light polymers; F, G, heavy polymers; ▆▆▆ fertilized; ▨▨▨ chemosterilized; ▦▦▦ unfertilized.

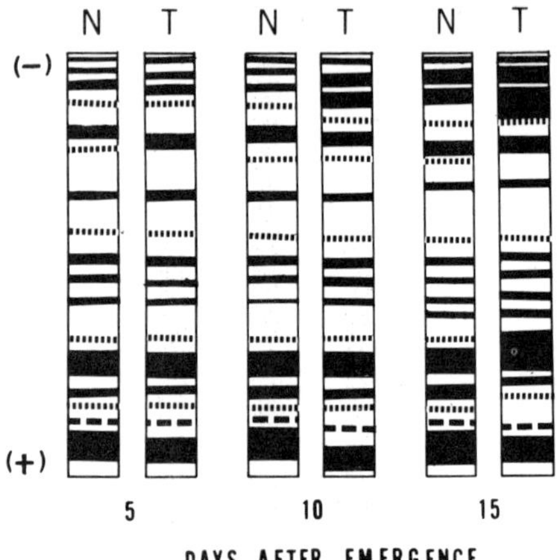

Fig. 11 Haemolymph proteins of normal (N) and chemosterilant-treated (T) female house flies after acrylamide gel electrophoresis (Gadallah, et al., 1972b. Reprinted by permission of the copyright owner).

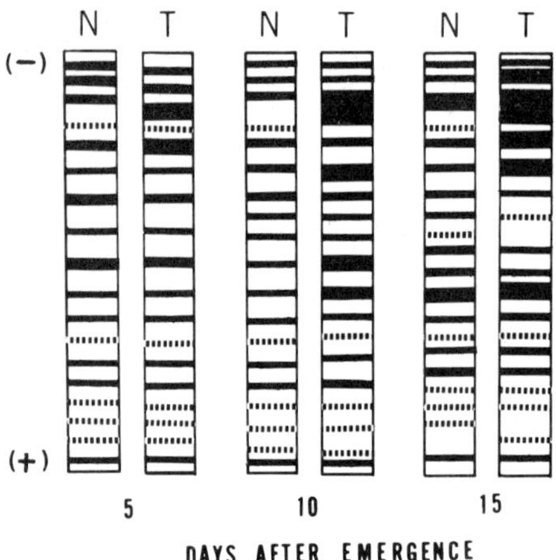

Fig. 12 Ovarian proteins of normal (N) and chemosterilant-treated (T) female house flies after acrylamide gel electrophoresis (Gadallah, et al., 1972b. Reprinted by permission of the copyright owner).

110

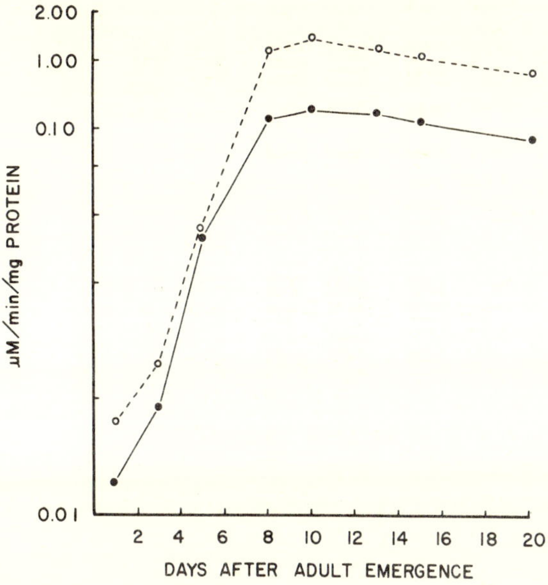

Fig. 13 Lactic acid dehydrogenase activity in the ovaries from normal and chemosterilized house flies. o——o from normal house flies; •——• from chemosterilized house flies (Gadallah, et al., 1972c. Reprinted by permission of the copyright owner).

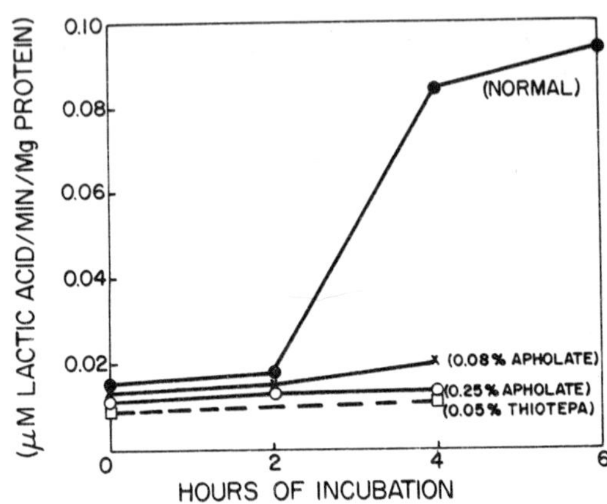

Fig. 14   Lactic acid dehydrogenase activity in
the eggs of normal and chemosterilized house flies.

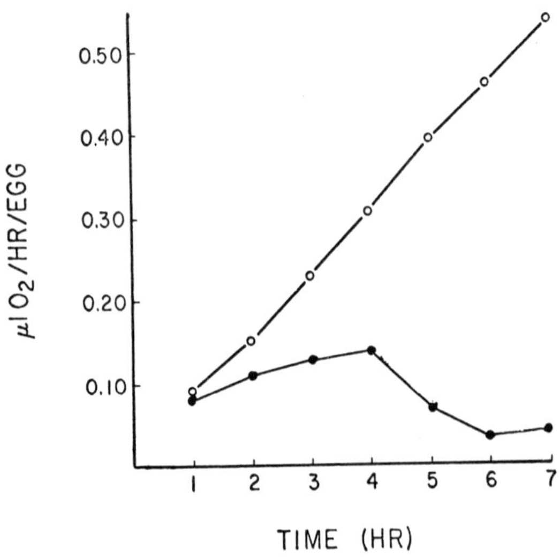

Fig. 15   Respiration rate of house fly eggs.
o——o fertilized; ●——● unfertilized.

SENESCENCE AND THE REGULATION OF CATALASE
ACTIVITY AND THE EFFECT OF HYDROGEN
PEROXIDE ON NUCLEIC ACIDS*

H. V. Samis, M. B. Baird, and H. R. Massie

Masonic Medical Research Laboratory
Utica, N. Y. 13501

I have chosen to take this opportunity to discuss some work which Dr. Massie, Dr. Baird and I are
doing with catalase (EC 1.11.1.6) and its primary
substrate, hydrogen peroxide.

Catalase has fascinated me for some time and for
a number of reasons. As you well know, catalase is a
very large, polymeric enzyme, with a hematin prosthetic group, and having a molecular weight of about
250,000. It is also ubiquitous in aerobic cells containing a cytochrome system (Deisseroth and Dounce,
1970). Catalase is generally found in subcellular
particles, the microbodies or peroxisomes, which also
contain urate oxidase, D-amino acid oxidase, xanthine
oxidase, cytochrome oxidase and α-hydroxy acid oxidase.

The activity of catalase can be irreversibly
destroyed in some cells with the herbicide 3-amino-
1,2,4,-triazole without affecting the resynthesis of
the enzyme (Heim, Appleman and Pyfrom, 1955). In
some respects, however, the most curious and tanta-
lizing characteristic of this enzyme is the myster-
iousness of its biological role or roles. We must,
nevertheless, reasonably assume that whatever the
physiological roles of catalase may be, they must in
some way involve either the catalatic or peroxidatic

---

*Research supported, in part, by a grant-in-aid
from the W. Alton Jones Foundation.

destruction of hydrogen peroxide.

Catalase activity in rodents is greatest in liver, kidney, and blood, as shown for CFN male rats in Fig. 1. Although we have detected low levels of catalase activity in cardiac and skeletal muscle as well as in brain, these levels are probably compatible with residual blood catalase activity contaminating our preparations.

The well known destructive properties of hydrogen peroxide, which I shall discuss later, and the ubiquitous distribution of catalase in aerobic cells makes this enzyme, in our view, an interesting one for study in the context of the regulation of enzyme activity in animals as they undergo senescence.

We have used the herbicide, 3-amino-1,2,4,-triazole ("AT") to destroy hepatic and renal catalase activity in male C57BL/6J mice of different ages, and then measured the renewal of activity over a twenty-four hour period. In these experiments the young (Y) animals were 312 days of age, the middle aged animals (M) were 616-646 days of age, and the old (O) were 955-1020 days of age. The results of these experiments may be seen in Table 1, where the catalase activities are expressed on a DNA base, and in Table 2 where they are expressed on a unit wet weight base.

We found that the initial levels of catalase activity were lower in the old animals than in the young (Tables 1 and 2). Two hours after injection with ("AT"), at a level of 1 mg per gram body weight, catalase activities dropped to approximately 10% of the initial values. The young and middle aged groups evidenced renewal of activity at approximately the same rate, whereas renewal of catalase activity was markedly slower amongst animals in the old age group. The relative hepatic and renal catalase activities over this 24 hour period after treatment with 3-amino-1,2,4,-triazole for all three age groups may be seen in Fig. 2.

In examining these data we were impressed by the obvious differences in the capacity of the individ-

uals in each age group to renew activity of this
enzyme following treatment with the herbicide. We
consequently scored each animal in each age group for
every time point as to the significant renewal of
activity during the 24 hour period.

The animals were grouped in two catagories for
the 4 through 24 hour time periods. The catagories
were: (1) that fraction of the animals showing "marginal renewal"; and (2) the fraction showing "absolute renewal". The definitions of these catagories
are given in Table 3 in which results of this manipulation of the data are also shown.

Although the treatment of these data may seem a
bit bizzarre, it does point up rather dramatically
the individual characteristic of senescence, at least
in terms of the renewal of catalase activity following "AT" treatment - organism age individually, not
in groups.

You will note that all young and middle aged
mice showed "marginal renewal" of hepatic catalase
activity from 4 through 24 hours but only 40% of the
old animals could be scored in this category at 4
hours, 60% at 6 hours, 25% at 10 hours, 75% at 12
hours and 50% at 24 hours. A similar picture was
obtained from renewal of renal catalase activity
following "AT" treatment. In the "absolute renewal"
category the pattern obtained shows an even more
striking age-related segregation. It could be said
that each animal in any age group has a physiological
age as regards catalase renewal after "AT" treatment
which is unique to it. This is, in our judgement,
potentially a point of considerable significance in
experimental gerontology which is frequently ignored
in the statistical evaluation of data obtained from
comparisons of many properties of form and function
in groups of animals as they age.

Since the red blood cell is a common denominator
in mammals, all of which are, as far as I know,
aerobic, the catalase activity of blood is also of
interest to us. Recently we have been routinely
measuring blood catalase activities in rodents used

in our ongoing experiments and some of the results
of this survey may be seen in Fig. 3. These data
were obtained on blood taken from the abdominal
aortae of CFN male rats. Here we also included data
derived from animals considerably younger than those
used in our experiments with catalase renewal. First
it must be remembered that "AT" does not affect blood
catalase activity (Rechcigl, Price and Morris, 1962),
which is found only in the red blood cells (Deisseroth and Dounce, 1970). The data on this graph show
that blood catalase activity decreases markedly in
animals from 90 to approximately 200 days of age,
with no perceptible change in activity in animals,
thereafter, through 1000 days of age.

Let me now digress slightly in order to deal
with the results of another study concerning the substantive characteristics of catalase derived from
rodents of different ages. We felt it was important
to know if the catalase in young animals is different
from that in the old. Of course, here, we are concerned not with the levels of catalase activity from
tissue derived from donors of various ages, but with
the catalase as a substantive unit, in particular its
specific activity.

After preparing and enriching hepatic catalase
from CFN male rats we determined the total yield of
hepatic protein, units per mg protein in the final
enriched preparation, and the characteristic absorb-
ancy ratios, i.e. that between 407 nm and 276 nm
(Price et al., 1962).

The data in Table 4 show that total hepatic
catalase activity decreased with age for these
animals. The units of catalase per unit total
hepatic protein also exhibited a decrement with age.
A small decrease in specific activity in the enriched
preparation was also found but this was well compen-
sated by the differences in 407 nm to 276 nm ratios
observed for the preparations (Price et al., 1962).
These results show no discernible substantive age-
related differences in hepatic catalase from CFN male
rats, although age-related differences do exist in

levels of activity and renewal capacity for this enzyme.

For many reasons, not the least of which is the marked cost advantage, we have begun studies of catalase, its activity and regulation in <u>Drosophila</u>. Since very little is known concerning catalase in <u>Drosophila</u>, I should like to present briefly a bit of background in this regard.

The activity of catalase is linear with the amount of <u>Drosophila</u> homogenate under our assay conditions (Fig. 4), and shows a negative substrate concentration dependence from 0.1 to 4.0 molar H_2O_2 (Fig. 5). The negative substrate concentration dependence is, we think, due to degradation of the enzyme protein by the peroxide, though we have not yet shown this, unequivically, in terms of the specific types of damage to the enzyme produced by H_2O_2.

Unlike catalase from rodents, catalase derived from <u>D. melanogaster</u> shows a moderately sharp pH dependence at pH 7.2 (Fig. 6). Furthermore, the activity in males is significantly higher than in females, and it is not equally distributed amongst the three gross anatomical segments of these insects (Fig. 7).

<u>Drosophila</u> catalase activity also changes markedly during pre-adult development. As can be seen in Figure 8, catalase activity is very low in newly laid eggs, increases to a maximum in early (light) pupae, and subsequently drops during metamorphosis to almost half maximum activity.

After eclosion catalase activity remains fairly constant for about 4 weeks and then decreases to approximately half the 4-week level by the 10th week (Fig. 9).

Since we were also still working with renewal of catalase activity following treatment with 3-amino-1,2,4,-triazole in rodents, we decided to see if we could use the same manipulative approach with <u>Drosophila</u>. We found that feeding "AT" to 3 week and 1 week old male <u>Drosophila melanogaster</u> in the aqueous portion of the media (Carolina Instant Media) for

12 hours resulted in a decrease in catalase activity
to approximately 2% of the pre-treatment level (Fig.
10). When the adulterated diet was replaced by a
normal diet, activity returned to the pre-treatment
level in one week. These results were compatible
with results of our previous experiments that showed
no difference in _Drosophila_ catalase activity until
after the fourth week.

When we compared renewal of catalase activity
following "AT" treatment on flies 1 and 7 weeks of
age, we found a clear age-related difference. The
old flies renewed activity much more slowly than did
the young. After 144 hours the 1 week old flies had
completely regained their pre-treatment level of
activity whereas the 7 week old flies had reached
only about half their pre-treatment activity level
(Fig. 11).

While these experiments were being carried out,
we were also looking at the effects of hydrogen
peroxide on _Drosophila_ survival. In one of our exploratory excursions in this area we placed one-,
three-, five-, and seven-week old male _D. melanogas-
ter_ on each of three standard preparations of media
made up with either water, 0.1 M H_2O_2 or 1.0 M H_2O_2
for 5 days and then scored the survivors each day
(Table 5). These are, of course, extremely high
levels of H_2O_2 which would not ordinarily be present
in the fly's diet. Nevertheless, we wanted to know
if we could impose a stress which would result in a
change in survival probability which might reflect
the age-related decrease in catalase activity we had
previously observed. The data in Table 5 show that,
after 24 hours, 1.0 M H_2O_2 in the diet of one-week
and three-week old flies has little or no effect on
survival, whereas 82.8% of the five-week old and
88.5% of the seven-week old _Drosophila_ died. Comparison with control populations shows the decrease
in survival with age is real. On the medium prepared
with 0.1 M H_2O_2 none of the one- and three-week old
flies died in the first 24 hours, but 10.8% of the
five-week old animals died as did 44% of the seven-

118

week old animals. These deaths were, however, no greater than those observed in a control population maintained on medium prepared with water. At 48 hours of age the young flies (one and three weeks old) were essentially unaffected by the 0.1 M H_2O_2 but 14% of the five week old flies and 49% of the seven week old flies died.

We also compared survival of male $\underline{D}$. $\underline{melanogaster}$ maintained from eclosion on medium with and without "AT" in the presence and absence of 0.01 and 0.1 M H_2O_2 on 0.01 and 0.1 M hydroxalamine.

The results of these studies can be seen in Table 6 where the percentile ranking for 80, 50, 30, and 10 percent survival are shown. For example, of the flies kept on media made with 0.01 M H_2O_2, 80% were still alive at 47 days, 50% were alive at 61 days and so on.

It is clear from these data that 0.01 M H_2O_2 in the absence of the herbicide has no effect on the median survival time nor on the general shape of the curve indicating the extreme 10% rank. When 3-amino-1,2,4,-triazole is included in the dietary water along with the 0.01 M hydrogen peroxide, a marked drop results in all ranks. This pattern is also seen for the population maintained on 0.1 M H_2O_2 in the presence of "AT". In the case of 0.1 M H_2O_2 in the absence of "AT" we also found a modest decrease in all ranks. Comparing these data we concluded that catalase can serve in a protective role in $\underline{Drosophila}$ as regards dietarily supplied H_2O_2. It is also evident that feeding "AT" alone has a modest but real effect on survival in all ranks suggesting the possibility that under normal conditions catalase does confer some protection via either its catalatic or peroxidatic mode of activity.

Further evidence in this regard was obtained by feeding hydroxalamine. It has been shown that the deleterious effects of hydroxalamine with respect to nucleic acids are of two general types depending on its concentration (Freese, Freese, and Graham, 1966). At low concentrations (0.01 M or lower) the effects of NH_2OH on nucleic acids are oxygen dependent and catalase inhibited. This suggests, of course, that NH_2OH at low concentrations serves as a peroxide generator. At high concentrations (0.1 M and above) the effects of NH_2OH on nucleic acids are not oxygen dependent nor are they inhibited by catalase. In Table 6 we have, therefore, included data on survival of $\underline{D}$. $\underline{melanogaster}$ males which were maintained on diets made with 0.1 and 0.01 molar NH_2OH in the presence and absence of "AT". In the absence of the herbicide fly populations maintained on medium made with 0.01 M NH_2OH yielded essentially the same values for the 80, 50, and 30 percent ranks as did controls, but did show a slight reduction in the extreme, 10 percent, survival rank.

The presence of "AT" in combination with 0.01 M H_2O_2 produced striking decreases in all ranks.

The results of these studies with $\underline{D}$. $\underline{melanogaster}$ show clearly that decreased catalase activity can result in decreased survival - activity of this enzyme is of importance to these organisms with regards to their survival probability.

We have also been very interested in determining if the age-associated difference in catalase activity and regulatory capacity, and the apparently associated effects of H_2O_2 on survival, which I have discussed, could be linked as regard causality to some effect of the peroxide consequence on the cellular information system.

Hydrogen peroxide has been shown to have degradative effects on macromolecules (Rhaese and Freese, 1968; Rhaese, Freese and Melzer, 1968; Butler and Conway, 1950; Butler and Smith, 1950; Moroson and Alexander, 1961). We chose to look at these effects in order to establish criteria for exploratory investigations in senescing organisms.

If you subject DNA to H_2O_2 a dramatic decrease in intrinsic viscosity ensues. The decrease is due primarily to double strand scissions (Fig. 12) and the effect is dependent upon H_2O_2 concentration and temperature (Fig. 12 and Table 7). In the presence of H_2O_2 at 37°C we have also found a decrement in neutral sedimentation coefficients of heat denatured DNA indicating an accumulation of single strand scissions (Table 8).

Cross linking in DNA is also produced by H_2O_2 (Fig. 13), as is a loss of conjugated double bond character which is indicative of base destruction (Fig. 14). Noteworthy in Figure 14 are the differences in the rate of loss of ultraviolet absorbing material between native and heat denatured DNA and RNA.

Do these _in vitro_ effects of H_2O_2 on nucleic acids have any parallels in nucleic acids derived from aging organisms?

We think they may - at least we have found some changes in hepatic DNA of the same sorts produced by exposure to H_2O_2.

We have determined the intrinsic viscosity of DNA in hepatic nuclear lysates derived from a number of CFN male rats of ages ranging from 31 to 1058 days (Fig. 15). The intrinsic viscosity was found to exhibit a marked decrement with age, representing a change in double-strand molecular weight from approximately 250×10^6 to about 25×10^6 daltons (Fig. 15). As can be seen in Figure 15, these changes occur early in the life of the rat.

At this juncture we felt it was necessary to satisfy ourselves that the viscosity being determined in these nuclear lysates was due to the DNA present and not to associated protein or RNA. To this end we prepared hepatic lysates from a 706 day old CFN male rat and treated one third of the lysates, each, with pronase, RNase, and DNase. Neither pronase nor RNase treatment resulted in any decrement in viscosity whereas DNase treatment caused the viscosity of the lysate to drop to that of the solvent.

Another possible cause for the observed decrease in DNA molecular weight as determined by changes in intrinsic viscosity could be the presence of higher levels of DNase activity in lysates obtained from the older animals.

We decided that one way to protect the lysate - derived DNA was to carry out the entire preparative procedure as well as the viscosity measurements themselves in the presence of a large excess of sonicated calf thymus DNA (200 mg DNA/ml).

The calf thymus DNA was sonicated so that it would make no contribution to the viscosity of the lysate sample. We compared viscosity of lysates from 318, 706, and 946 day old rats prepared in the presence and absence of the sonicated DNA. The presence of the carrier DNA had no effect on the intrinsic viscosity of any of the lysates, regardless of the age of the donor rat.

A decrement, early in life, was also found in sedimentation coefficient of hepatic DNA in nuclear lysates from male CFN rats, representing a decrease in single-strand molecular weight from about 3.5×10^6 to approximately 1.0×10^6 daltons (Fig. 16).

Finally, we found that DNA from like sources showed an early age-related increase in the level of cross-linking (Table 9).

We, of course, cannot argue from the results of our _in vitro_ studies for a cause and effect relationship between them and the results of our _in vivo_ studies. We can say, however, that three of the four _in vitro_ changes produced by H_2O_2 are clearly detectable in hepatic DNA in the early days of the life of these animals.

We feel that much of what I have discussed today goes to support the notion that all the causes of senescent deterioration may not occur in the senescing organism, but rather in the early stages of growth and development. This is clearly the case for the changes in DNA which we have observed as well as for changes in blood catalase levels.

It is certainly apparent that the age-related changes in catalase activity and the capacity to renew catalase activity following its distruction with 3-amino-1,2,4,-triazole occurred late in the life of our experimental animals. This does not necessarily mean, however, that the primary causes for these changes did not precede them by many weeks, days, or months.

In my view, the primary question confronting experimental gerontology is what causes the senescent deterioration of biological form and function? We think that this element of causality is central to an understanding of the phenomenon though it may prove cumulative and even autocatalytic in nature. In our view it is not enough to show coincidence between change and age. The change must be of consequence to the vigor of the organism. That the color of hair or the activities of some enzymes show age-related changes may be nothing more than coincidental with advancing age, and of no consequence to the organism's order of form or level of function.

Of course, we may well stand in error, but nevertheless, in our view, a prejudice is widespread to the effect that the causes of senescence occur and will, therefore, be found in senescent organisms. This, obviously need not necessarily be so.

Thus it is quite possible that an organism is laid at irrevocable risk long before adulthood is reached, even as early as the time of embryogenesis.

REFERENCES

Alberts, B. (1967). _Biochemistry_ 6, 2527.

Baird, M.B. and Samis, H.V. (1971). _Gerontologia_ 17, 105.

Butler, J. and Conway, B. (1950). _J. Chem. Soc._ 3418.

Butler, J. and Smith, K. (1950). _Nature_ 165, 847.

Crothers, D. and Zimm, B. (1965). _J. Mol. Biol._ 12, 525.

Deisseroth, A. and Dounce, A.L. (1970). _Physiol. Rev._ 50, 319.

Freese, E., Freese, E.B. and Graham, S. (1966). _Biochim.Biophys_. Acta 123, 17.

Goldstein, D.B. (1968). _Anal. Biochem._ 24, 431.

Heim, W.G., Appleman, D. and Pyfrom, H.T. (1955). _Science_ 122, 693.

Moroson, H. and Alexander, P. (1961). _Radiation Res._

14, 29.
Pogo, A.O., Allfrey, V.G. and Mirsky, A.E. (1966).
    Proc. Nat. Acad. Sci. Wash. 56, 550.
Price, V.E., Sterling, W.R., Tarantola, V.A., Hartley,
    R.W., Jr., and Rechcigl, M., Jr. (1962). J. Biol.
    Chem. 237, 3468.
Rechcigl, M., Jr., Price, V.E. and Morris, H.P. (1962).
    Cancer Research 22, 874.
Rhaese, H. and Freese, E. (1968). Biochim. Biophys.
    Acta 155, 476.
Rhaese, H., Freese, E. and Melzer, M. (1968). Biochim.
    Biophys. Acta 155, 491.
Samis, H.V., Baird, M.B. and Massie, H.R. (1972).
    J. Insect Physiol. 18, 991.
Studier, F. (1965). J. Mol. Biol. 11, 373.
Zimm, B. and Crothers, D. (1962). Proc. Nat. Acad.
    Sci. Wash. 48, 905.

125

TABLE 1

Recovery of catalase activity in C57BL/ 6J male mice following intraperitoneal injections of 3-amino-1,2,4,-triazole at the level of 1 g/kg body weight. Activity is expressed as units per micromole deoxyribose apparent as DNA. Catalase activity was determined as described previously (Baird and Samis, 1971).

Organ	Age Group	Hours after injection							
		0	2	4	6	8	10	12	24
Liver	Y	13315± 3742[1]	2061± 665	2717± 527	4176± 2328	5433± 1488	7106± 2033	8787± 2227	9693± 1608
	M	13389± 3631	1238± 462	3798± 462	4463± 2328	3419± 832	4324± 1885	7891± 4565	10691± 4934
	O	9545± 1885[2]	1765± 979	1700± 684	2301± 1109	1850± 1423	1007± 702[2]	2827± 1460[2]	2486± 2319
Kidney	Y	1238± 65	194± 28	185± 37	213± 74	305± 74	453± 92	462± 92	628± 55
	M	1100± 46	111± 18	185± 65	231± 65	305± 37	416± 74	360± 92	591± 129
	O	767± 111	102± 28	102± 28	157± 46	231± 111	129± 102[2]	222± 102	240± 166

[1] Mean ± 1 SD.
[2] N = 4.

(From Baird and Samis, 1971, Gerontologia 17, 105, S. Karger, Basel)

TABLE 2

Recovery of catalase activity in C57BL/ 6J male mice following intraperitoneal injections of 3-amino-1,2,4,-triazole at the level of 1 g/kg body weight. Activity is expressed as units per gram wet weight of tissue. Catalase activity was determined as previously described (Baird and Samis, 1971).

Organ	Age Group	Hours after injection							
		0	2	4	6	8	10	12	24
Liver	Y	4597± 238[1]	594± 74	793± 72	1515± 298	2225± 303	2465± 642	2849± 178	2809± 231
	M	5002± 469	418± 69	1042± 391	1261± 312	1601± 220	1705± 439	2033± 785	2888± 691
	O	3930± 735	508± 148	541± 92	776± 312	941± 520	480± 252[2]	1144± 716[2]	811± 704[2]
Kidney	Y	1406± 145	215± 59	192± 23	251± 65	334± 58	400± 51	517± 67	833± 230
	M	1178± 165	130± 21	195± 63	232± 65	329± 51	396± 51	419± 83	631± 97
	O	758± 176	118± 28	108± 18	170± 65	244± 117	131± 79[2]	267± 114[2]	295± 675[2]

[1] Mean ± 1 SD.
[2] N = 4.

(From Baird and Samis, 1971, Gerontologia 17, 105, S. Karger, Basel)

TABLE 3

Frequency of individual mice which renew catalase (as units/ g wet wt.) at various periods of time following injection of 3-amino-1,2,4,-triazole (1 g/kg body wt.)

Organ	Hours after 3-AT injection	Marginal Renewal[1]			Absolute Renewal[2]		
		Y	M	O	Y	M	O
Liver	4	1.00	1.00	0.40	0.00	0.20	0.00
	6	1.00	1.00	0.60	1.00	0.60	0.20
	8	1.00	1.00	0.60	1.00	1.00	0.40
	10	1.00	1.00	0.25*	1.00	0.80	0.00*
	12	1.00	1.00	0.75*	1.00	0.80	0.50*
	24	1.00	1.00	0.50*	1.00	1.00	0.25*
Kidney	4	0.60	0.80	0.00	0.00	0.00	0.00
	6	1.00	0.80	0.40	0.00	0.00	0.00
	8	1.00	1.00	0.80	0.20	0.20	0.00
	10	1.00	1.00	0.25*	0.60	0.80	0.00*
	12	1.00	1.00	0.75*	1.00	0.80	0.25*
	24	1.00	1.00	0.50*	1.00	1.00	0.50*

[1]Activity > mean activity (t + 2h) + 2SD.
[2]Activity > 2 x (mean activity t + 2h + 2SD)
*N = 4
(From Baird and Samis, 1971, _Gerontologia_ 17, 105, S. Karger, Basel.)

TABLE 4

Characteristics of CFN male rat hepatic catalase enriched by purification according to the method of Price et al. (1962). Catalase activity was determined as previously described (Baird and Samis, 1971).

Age # in days	Total hepatic catalase activity (units)	units / mg protein in W.H.	units / mg protein in the enriched preparation	O.D. 407nm/ 276 nm*
205	332,000	178.01	17,118	1.110
511	346,000	146.85	16,513	1.095
849	139,000	127.88	14,931	1.044

*The major contaminant in such preparations is ferritin. A contamination of 1.0% ferritin lowers the O.D. 407:O.D. 276 to 1.02. Thus, these preparations contain less than 1.0% ferritin as judged by spectral data.

TABLE 5

Age in Weeks

Hours	1 Week			3 Weeks			5 Weeks			7 Weeks		
	H_2O	$0.1M\ H_2O_2$	$1.0M\ H_2O_2$	H_2O	$0.1M\ H_2O_2$	$1.0M\ H_2O_2$	H_2O	$0.1M\ H_2O_2$	$1.0M\ H_2O_2$	H_2O	$0.1M\ H_2O_2$	$1.0M\ H_2O_2$
24	100	100	98	99.2	100	99.2	92.8	89.2	17.2	56.7	56.0	11.5
48	100	100	52.8	98.4	98.8	24.0	79.2	86.0	2.8	53.3	51.5	2.5
72	100	100	20.4	92.6	98.8	3.2	76.8	73.2	0.4	33.9	37.0	1.2
96	100	99.6	1.2	97.6	98.8	0	75.2	68.8	0	32.7	32.0	0
120	100	99.2	0	96.8	97.2	0	70.4	61.2	0	29.3	18.0	0

Survival of _Drosophila melanogaster_ males
during 5 day treatment with hydrogen peroxide in the diet.

TABLE 6

Percentile Ranking of Survival Time
Drosophila melanogaster Ore-R. Wild Type Males

% Survival	Days									
	0.1% AT	0.1M NH_2OH + 0.1% AT	0.1M NH_2OH	0.01M NH_2OH + 0.1% AT	0.01M NH_2OH	0.1M H_2O_2 + 0.1% AT	0.1M H_2O_2	0.01M H_2O_2 + 0.1% AT	0.01M H_2O_2	H_2O
80	31	3		23	43	4	33	22	47	42
50	47	3	3	29	56	5	38	28	61	57
30	54	5		35	61	6	43	31	66	63
10	65	7	13	43	66	8	51	34	72	72

TABLE 7

Effect of temperature and H_2O_2 concentration on the rate of DNA degradation in SSC buffer. DNA concentration was 190 µg/ ml.

Temperature, °C	H_2O_2,M	$t_{1/2}$, hrs.
25	0.088	11.5
37	0.088	5.6
37	0.000	0.0
37	0.0088	17.5
37	0.00088	203.0
45	0.088	0.82

$t_{1/2}$ = time required for the molecular weight to decrease by one half.

TABLE 8

Neutral sedimentation coefficients for heat denatured calf thymus DNA treated with 0.088M H_2O_2 at 37°C for different periods of time.

Time (hours)	$S_{w, 20}$
0	21.9
0.5	19.1
1	17.9
3	12.6
6	9.34

TABLE 9

Percent of hepatic DNA cross-linked in CFN male rats
of different ages. Cross-links were determined using
the method of Alberts (1967).

Age, Days	n	Cross-Links
71-147	9	1.472 ± 0.180
249-283	6	1.490 ± 0.170
403-526	7	3.443 ± 0.124
616-798	5	2.948 ± 0.304
846-946	3	3.213 ± 0.303

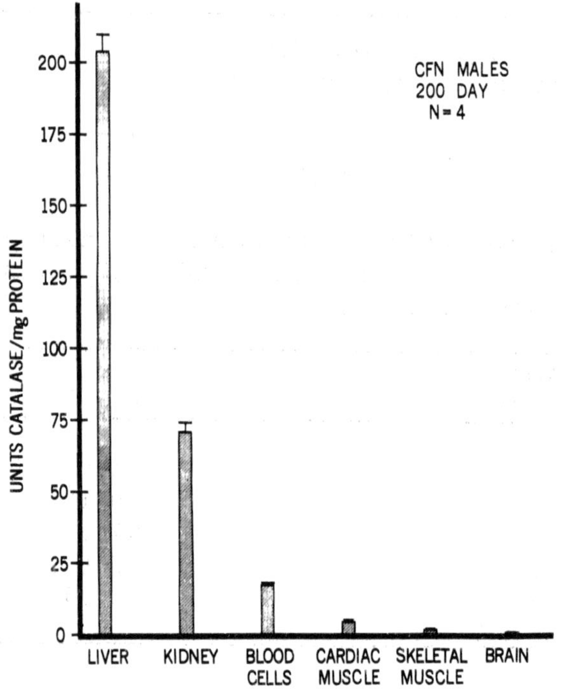

Fig. 1 Distribution of catalase activity in
CFN male rats. Catalase activity was determined as
described previously (Baird and Samis, 1971).

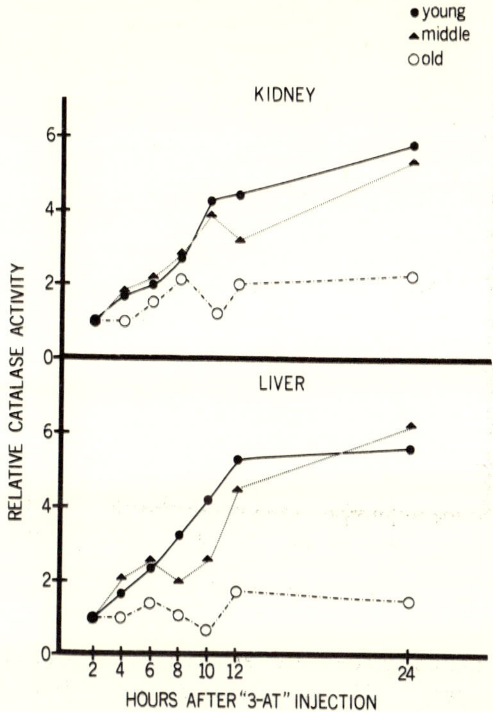

Fig. 2 Recovery of catalase activity in C57BL/6J male mice following intraperitoneal injection with 3-amino-1,2,4,-triazole. Activity is based on units catalase per micromole deoxyribose apparent relative to the activity two hours after drug treatment. (From Baird and Samis, 1971, *Gerontologia*, 17, 105, S. Karger, Basel.)

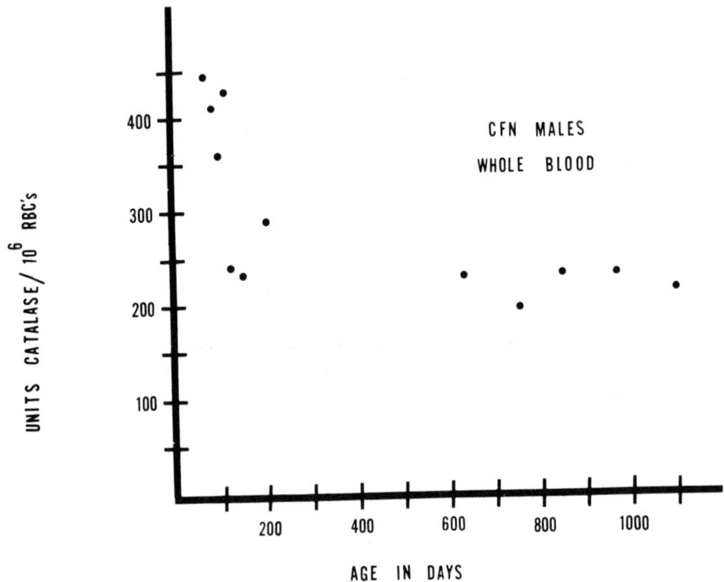

Fig. 3   Blood catalase activity in CFN male rats of different ages.   Blood samples were obtained from the abdominal aorta.   Red blood cell counts were made in a hemocytometer.   Catalase activity was determined in whole blood diluted in distilled water to yield lysed red blood cells.   Enzyme assays were determined as previously described (Baird and Samis, 1971).

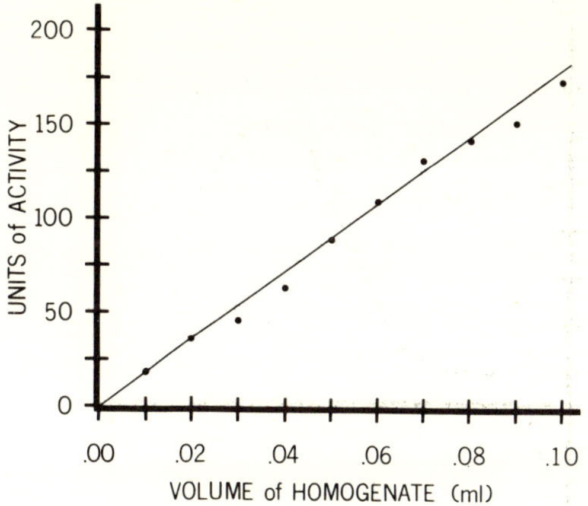

Fig. 4 The effect of enzyme concentration on catalase activity in <u>D. melanogaster</u>. Activity expressed as observed catalase activity per ml homogenate. (From Samis, Baird, and Massie, 1972, <u>J. Insect Physiol</u>. <u>18</u>, 991, Pergamon Press.)

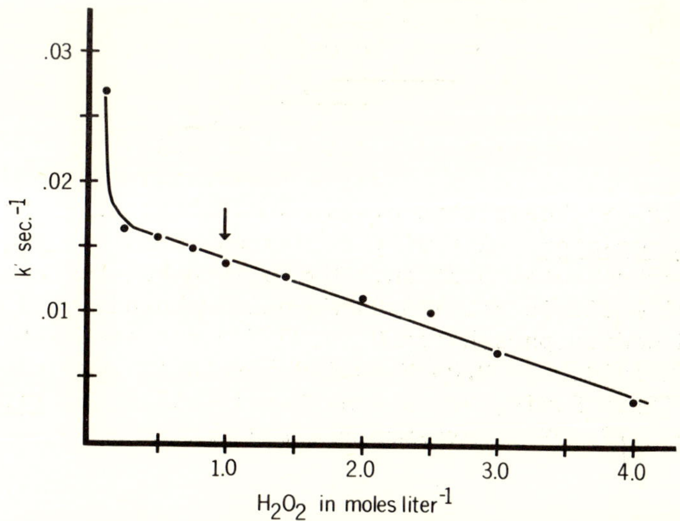

Fig. 5 The effect of substrate concentration on catalase activity in <u>D. melanogaster</u>. Activity is expressed as k', the first-order rate constant for the destruction of substrate. The arrow represents the concentration of substrate (0.98 M) normally used in this assay. (From Samis, Baird, and Massie, 1972, <u>J. Insect Physiol</u>. <u>18</u>, 991, Pergamon Press.)

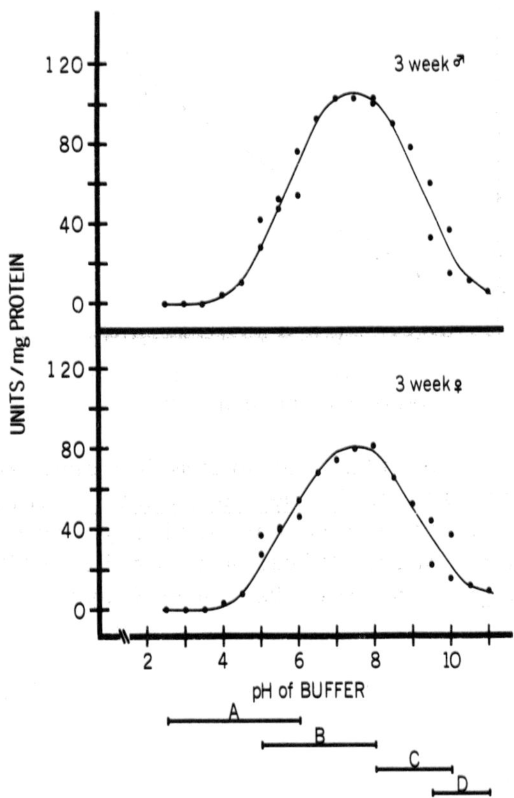

Fig. 6   The effect of pH on catalase activity in
D. melanogaster.   Activity expressed as units catalase per mg total protein.   Composition of buffers:
A, 0.05 M citric acid--0.05 M sodium phosphate; B,
0.05 M sodium phosphate; C, 0.05 M boric acid-potas-
sium chloride; D, 0.05 M sodium borate-sodium hydrox-
ide.   (From Samis, Baird, and Massie, 1972, J. Insect
Physiol. 18, 991, Pergamon Press.)

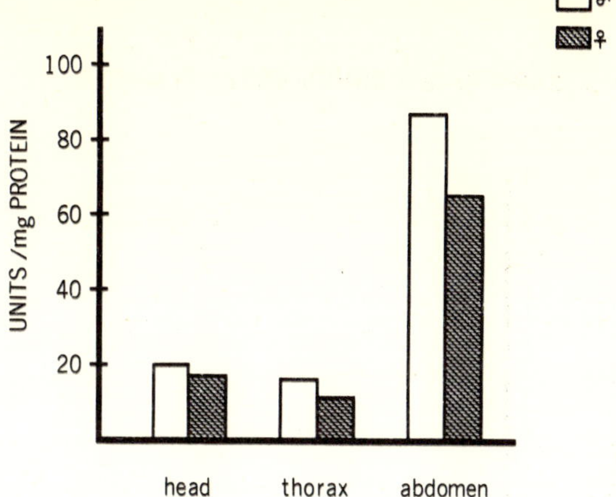

Fig. 7 Distribution of catalase activity in
Drosophila. Enzyme activity is expressed in units/
mg protein. (From Samis, Baird, and Massie, 1972.
_J_. _Insect_ _Physiol_. _18_, 991, Pergamon Press.)

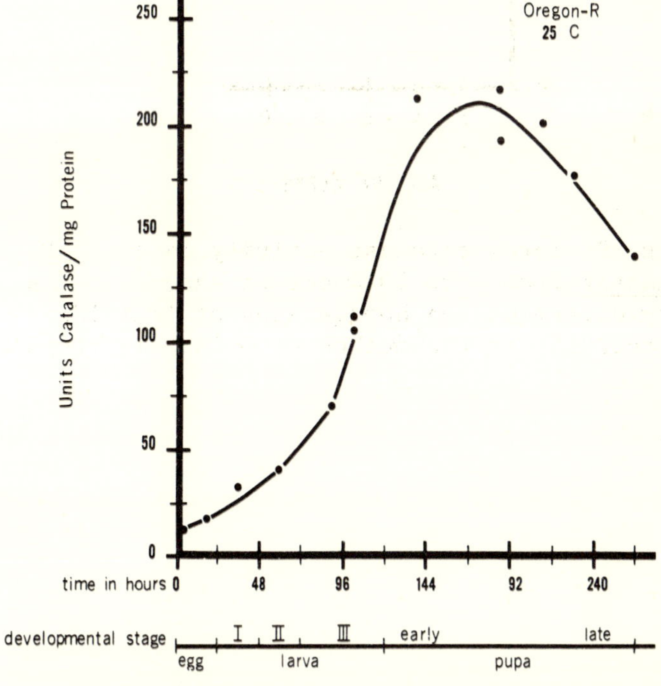

Fig. 8 Catalase activity during preadult development in _D_. _melanogaster_. Catalase activity was
determined using the YSI oxygen monitor as described
by Goldstein (1968). One unit of catalase activity
equals that amount of enzyme which yield 1 μ mole of
O_2 per minute at 30°C and are expressed as μ moles
oxygen evolved per minute.

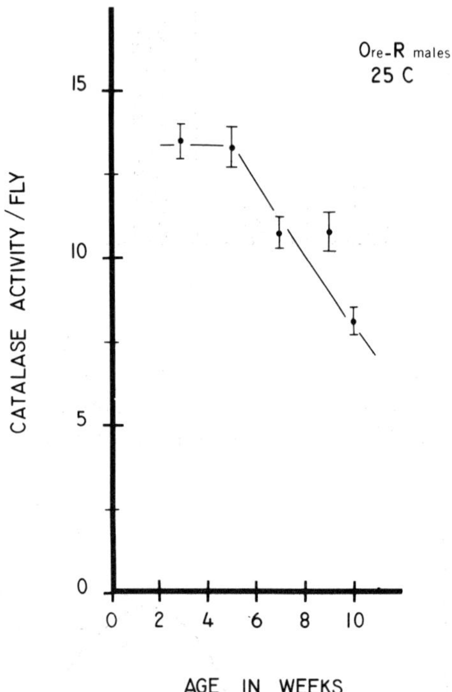

Fig. 9 Total catalase activity in adult <u>D. melanogaster</u> from 2 to 10 weeks of age. Catalase activity determined on homogenates of individual flies according to the method of Goldstein (1968).

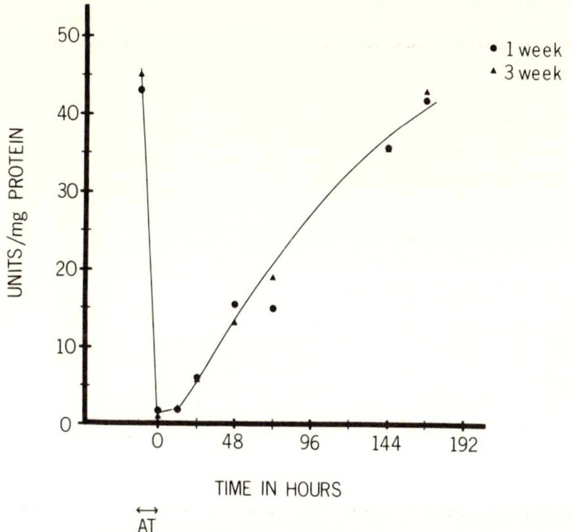

Fig. 10 Renewal of catalase activity in D. melanogaster males following a single 12 hour feeding on medium containing 3-amino-1,2,4,-triazole. Initial points (time 0 minus 12 hours) represent catalase activity prior to treatment. Activity expressed as units catalase per mg total protein. (From Samis, Baird, and Massie, 1972, J. Insect Physiol. 18, 991, Pergamon Press.)

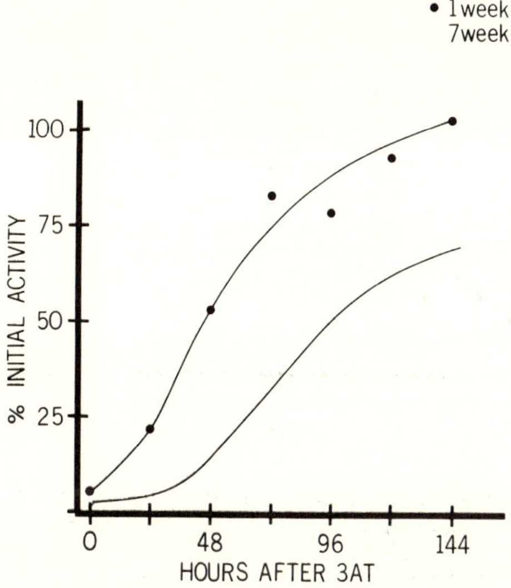

Fig. 11 Renewal of catalase activity in male D. melanogaster of different ages following treatment with the herbicide 3-amino-1,2,4,-triazole. Catalase assays and treatment of the flies with AT was essentially as previously described.

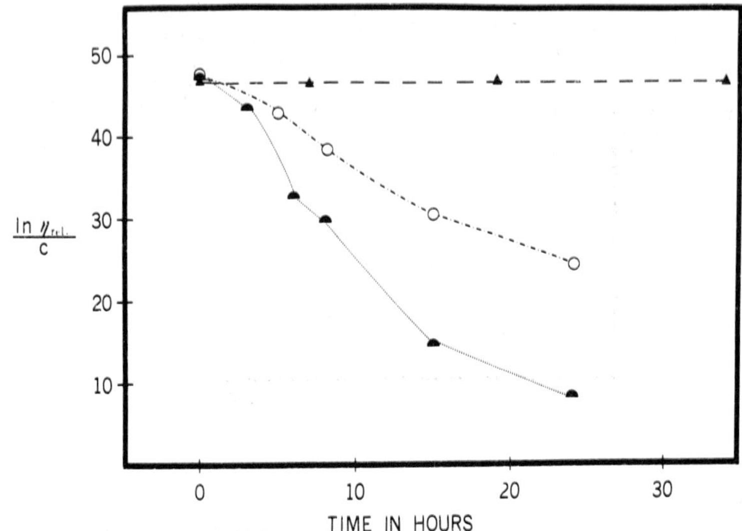

Fig. 12   Effect of H_2O_2 on the intrinsic viscosity of solutions of calf thymus DNA.   DNA concentration was 190 µg / ml.   ▲ - DNA in SSC.   ◓ - DNA in SSC and 0.088M H_2O_2.   o - DNA in SSC and 0.0088M H_2O_2.

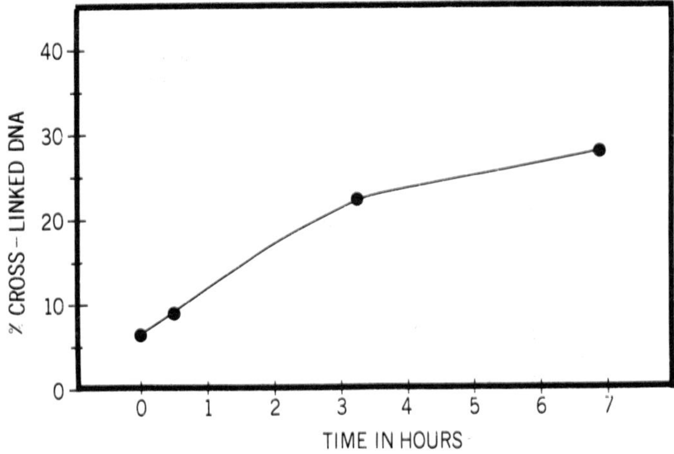

Fig. 13   Percent cross-links produced in calf thymus DNA by treatment with 0.088M H_2O_2 at 37ºC. Assay for cross-linking was carried by partitioning in the dextran-polyethylene glycol system as described by Alberts (1967).

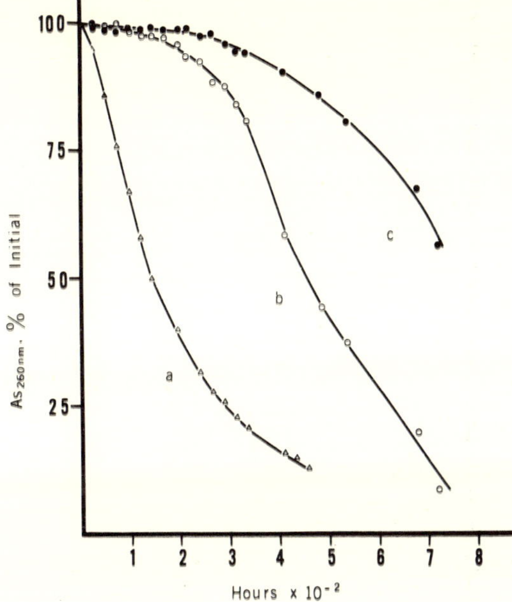

Fig. 14   Loss of ultraviolet absorbance (260 nm) in solutions of DNA and RNA in the presence of 0.088M H_2O_2 at 37°C in SSC.   a. RNA.   b. DNA denatured by heating to 100°C.   c. Native calf thymus DNA. Initial absorbance for all samples was 2.0 at 260 nm.   Absorbance from 360 to 220 nm was determined using a Cary Model 11 M recording spectrophotometer.

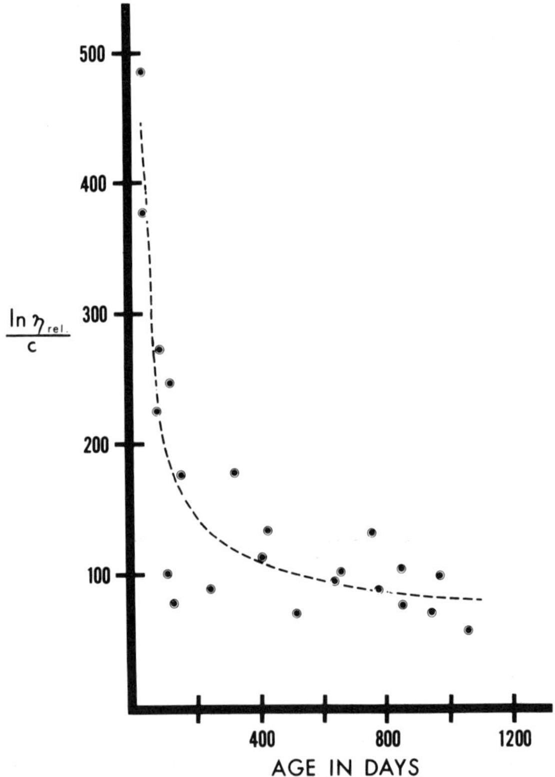

Fig. 15 Intrinsic viscosity of hepatic DNA in nuclear lysates from CFN male rats of different ages. Nuclei were isolated essentially as described by Pogo et al., (1966), suspended in 0.15 M NaCl, 0.015 M sodium citrate (SSC) and lysed in 1.0% Sarkosyl. The specific viscosity of the lysates were determined in a Zimm-Crothers viscometer (Zimm and Crothers, 1962) at a sheer stress of 0.0045 dynes / cm^2 at 25° C. At the DNA concentrations used and under these sheer conditions the intrinsic viscosities [η] is, for all practical purposes, equal to the natural log of the relative viscosity divided by the concentration (Crother and Zimm, 1965).

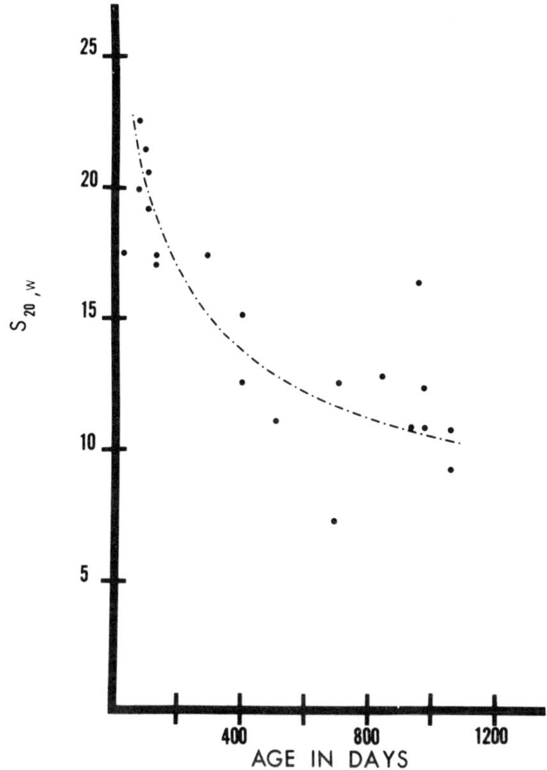

Fig. 16   Sedimentation velocities of hepatic DNA
in nuclear lysates from CFN male rats of different
ages.   Sedimentation velocities were determined at
40,000 rpm in a Model E analytical ultracentrifuge
using techniques described by Studier, (1965).
Nuclear lysates were assayed in 12 mm cells at a DNA
concentration of 20 µg/ml at 20°C, in alkaline solvent (0.1 M NaOH, 0.9 M NaCl).

HORMONAL REGULATION OF PROTEIN
SYNTHESIS IN INSECTS

Narayan G. Patel

Central Research Department
Experimental Station
E. I. du Pont de Nemours and Co.
Wilmington, Delaware 19898

Mr. Chairman, Ladies and Gentlemen:

To a biologist, who is interested in the various
aspects of growth and development and is seeking to
understand their molecular mechanisms, insects offer
a tailor-made design. One has a tremendous choice,
and can elect for his endeavors any one of more than
650,000 species.

In addition to the superlative varieties, insects lie between bacteria and man. Its biomass and
organismic complexity are quite convenient for us to
handle, manipulate, and analyze. Also, many insect
species are economically important and disease vec-
tors. Accordingly, desired protection of agricultur-
al products and guarding the health of humans and
domestic animals continue as a costly enterprise.
Finding selective ways to combat only the pest insects using strategies that would least upset the
balance of nature is in itself a tall order. Their
study, therefore, not only provides fulfillment for
some scientists but justification as well as support
of research projects.

Of all their features, the one which is rather
unique to insects is 'metamorphosis'. The changes in
the size and shape of the body which occur with pre-
dictable precision and temporal accuracy, through egg,
larval, and pupal development, have evolved to take

maximum advantage of the surrounding environment and
to withstand the stresses of hostile conditions.

The female lays eggs near the food sources and
the hatched larvae devour the food and grow continuously without much resistance offered by their soft
body walls. Final metamorphosis ensues at the termination of this growth phase. The resultant quiescent (pupal) stage undergoes profound biochemical as
well as morphological changes to emerge as a winged
adult. The well-differentiated adult form is cap-
able of flight and reproduction and suitably adopted
to encompass a much wider territory. I have viewed
this basic biology in 3 overlapping phases of growth,
development with differentiation, and reproduction.
I have elected to discuss here their control and
regulation by hormones, and specifically to share
with you our findings on the protein and RNA synthesis as affected by juvenile hormone (JH) and ecdysone (prothoracic gland hormone or PGH).

As early as 1917, Kopeć demonstrated that when
ligated near the thorax, _Lymantria_ larvae failed to
complete transformation to the subsequent pupal
stage. He envisaged the role of brain in this phenomenon. Subsequent work by others (see Novak, 1965
and Wigglesworth, 1970 for historical background)
indicated that a hormonal factor was involved. Thus,
some 17 years later Wigglesworth (1934, 1935, 1936)
showed in a series of classical and pioneering experiments with _Rhodnius_, that the brain produced a
hormone that regulated molting in insects. His subsequent studies clarified the concepts of insect
endocrinology and he postulated the following mechanism.

"The brain at the back of the head contains
'neurosecretory cells' the 'axons' from which end in
the 'corpus cardiacum' where the brain hormone is
set free into the blood. This acts on the 'thoracic
gland' and causes this to secrete 'molting hormone'.
When the molting hormone alone acts on the 'epidermal cells', these cells during molting produce an
adult type cuticle. But when the molting hormone

acts in the presence of 'juvenile hormone' secreted by the 'corpus allatum' lying just behind the corpus cardiacum, the epidermal cells produce a larval type cuticle." With very slight modifications, this concept presently holds for epidermal cells of virtually all insects.

The second major milestone was laid by two important later discoveries; the isolation of molting hormone 'ecdysone' by Butenandt and Karlson (1954) and preparation of a lipoid extract with juvenile hormone activity by Williams (1956).

Ecdysone could induce pupation in the isolated ligated posterior half of the Calliphora last larval stage, which otherwise was prevented from molting into pupa. As the molting involves processes concerned with cuticle formation and differentiation of cells, the concepts developed by Karlson and Sekeris (1966) centered around the regulation of genes and specifically ecdysone binding to DNA-dependent RNA and affecting a specific RNA polymerase for the synthesis of dopa-decarboxylase, which is necessary for puparium formation. Ecdysone (Clever and Karlson, 1960; Clever, 1963) and JH (Laufer and Holt, 1970) also induce 'puffing' in the giant polytene chromosomes of Chironomus. This is presented as the main argument for the specificity of the gene action; however, some serious questions have been raised since 'puffing' could likewise be induced by non-specific agents, e.g., ions, pH changes, etc. (Kroeger, 1963, 1968; Kroeger and Lezzi, 1966).

Application of the lipoidal extract prepared by Williams (1958) inhibited differentiative processes, maintained 'status quo' and caused the lepidopterous pupae to molt into a second pupa. Topically applied in a 'wax' medium to a localized area prior to the molt, only a small patch of cells remain undifferentiated while the rest of the body molts into the advanced stage (Schneiderman and Gilbert, 1957). Based on their studies with holometabolous insects, these workers developed a concept essentially similar to that of Wigglesworth based on studies of Rhodnius.

However, they modified Wigglesworth's original concept to include the slight effect of JH and somewhat
greater effect of prothoracic gland hormone (PGH or
ecdysone) on the chromosomes. This, in turn, affected RNA synthesis and, ultimately, the type of cuticle
made. They also indicated that the relative amounts
of JH and PGH at any given time may be responsible
for the synthesis of a larval, pupal or adult type
cuticle. The molecular mechanisms of JH action was
also investigated by Ilan et al. (1970). Using
Tenebrio pupae, they demonstrated that JH regulated
translation mechanisms affecting RNA dependent protein synthesis.

I shall expound more on both of these concepts
on the action of JH and ecdysone in the latter part
of my discussion. Let me briefly list the mimics
and analogues used in our investigations. They
include 'JH', used as a generic name; DME, dodecyl
methyl ether; FME, farnesyl methyl ether, (Schneiderman et al., 1965); LJM, Law's JH Mixture (Law et al.,
1966); CJH, Cecropia JH (Röller et al., 1967); BGE,
Bower's geraniol ether, methylene dioxyphenyl 6,7
epoxy geraniol ether (Bowers, 1969); MH, molting
hormone, used as a generic name for ecdysial-agents;
including ecdysone, ecdysterone and inokosterone as
assayed by Krishnakumaran and Schneiderman (1968)
(also referred to by Berkoff, 1969). The extensive
synthesis carried out both to elucidate the structures of insect hormones and to prepare compounds
which will have activities similar to native hormones
are discussed in various reviews (Berkoff, 1969;
Gilbert, 1969).

For the present discussion as a very broad and
generalized concept, the mechanism of JH is postulated to maintain the "status quo" by inhibiting cellular processes of differentiation while MH is thought
to activate the machinery that is essentially linked
to the ecdysial processes, prompting and preparing
epidermal cells for ensuing events of differentiation. It is crucial to know which is the primary
event and what other processes are affected during

the elucidation of the hormone action.

I would now like to discuss the macromolecular synthesis occuring concomitant with the developmental and metamorphic changes, during larval-larval (L-L) growth, larval-pupal-adult (L-P-A) development and adult-egg reproduction.

GROWTH

The growth exclusive of pronounced differentiation of the immature larval stages of insects is attained by periodic ecdysis of the larvel integument. Imaginal structures of ectodermal origin like the wing disc grow but do not molt. This feature could provide interesting and valuable information as these cells have the capacity to grow but not to differentiate. We examined the changes in the capacity of these wing discs of the silkworm, Samia cynthia ricini, to synthesize RNA and protein during L-L and L-P apolyses under normal development and hormonal influence (Patel and Madhavan, 1969). The hormones administered included ecdysone, ecdysterone, DME and LJM.

Materials and Methods

Larvae of the non-diapausing silkworm, S. cynthia ricini, were used. Their life cycle and methods are described (Schneiderman, 1967). In the present experiments, III, IV and V larval instars lasted approximately 5, 6 and 9 days, respectively. Near the end of V instar, the larvae spun cocoons and pupated 4 days later. The experiments were performed on III, IV, and V instar larvae and pharate pupae of specific ages. Three 5-day-old IV instar larvae were injected with specific hormone preparations. The controls received appropriate amounts of the carrier solvents. Twelve hours following treatment, the discs were dissected out and incubated at 25°C for 4 hrs. in insect Ringer (7.5 g NaCl, 0.35 g KCl, 0.2 g $CaCl_2$ in 1 liter water) with ^3H uridine and

149

^{14}C amino acid precursors for RNA and protein syntheses, respectively. The incorporation was terminated
by adding 200 µl 10% cold perchloric acid (PCA).
Other experimental details are reported by Patel and
Madhavan (1969) and include procedures for drying,
weighing, double label scintillation counting, etc.,
employed in this study.

Results

The imaginal wing discs are four small flatten-
ed invaginated epithelial pouches, one on each side
of the meso- and metathorax just internal to the
epidermis. The changes in the dry weight of the
wing disc from III instar to pupation are shown in
Fig. 1. The wing disc increases in dry weight from
8 µg in late II instar to 300 µg at the end of the
pharate pupal period. The growth rate of the disc
tissue continues to increase progressively up to the
pharate pupal stage, followed by an abrupt increase
marking the onset of wing eversion. By the end of
the eversion, the wings are drawn out of the body to
lie beneath the loose cuticle.
The growth of the tracheal system in the interecdysial period deserves comment. In Rhodnius, the
tracheal system is lined by an unbroken cuticular
membrane and no new tracheae or tracheoles appear to
be formed during the interecdysial period (Wigglesworth, 1954; Locke, 1964). This is not true in the
case of the Ricinni silkworm, where the tracheae of
larval imaginal discs grow extensively by branching
during the interecdysial period. Indeed, they grow
extensively over the terminal plastic window of the
isolated abdomen of diapausing pupae which have no
prothoracic glands (A.K. Krishnakumaran, unpublished). Apparently, the growth of the tracheal system
is not closely linked to molting in Lepidoptera
(Novak, 1965) as in Rhodnius.
Fig. 2 shows the titer of extractable JH during
the larval instars and pharate pupal period (A. K.
Krishnakumaran and H.A. Schneiderman, unpublished).

It also records estimates of ecdysone titers of the commercial silkworm <u>Bombyx</u> <u>mori</u> (Shaaya and Karlson, 1965). The JH concentration is highest immediately following each ecdysis and decreases rapidly by the middle of the instar. Its concentration continues to decrease at a slow rate during the rest of the instar and again increases just before the ecdysis, to reach the maximum immediately following the ecdysis.

Furthermore, the maximum and minimum JH titer in III instar are higher than the maxima and minima reached in IV and V instars. Also, the III instar contained more JH per g body wt. than did the succeeding larval instars and pharate pupa. At the onset of spinning, the concentration of JH falls to a low level and remains there until pupation. The ecdysone concentration is almost zero immediately after each ecdysis but begins to increase during the middle of each succeeding instar, reaching a maximum just before ecdysis. In general, the pattern of ecdysone concentration in larval and pharate pupal stages is approximately reciprocal to that of JH.

To determine how these changes in the hormonal milieu might affect RNA and protein syntheses, the capactities of the wing discs to incorporate radioactive precursors of RNA and protein <u>in</u> <u>vitro</u> were measured after their extirpation from larval and pharate pupal stages. The results recorded in Fig.2 reveal that ^{3}H uridine incorporation increases approximately at ecdysis and gradually decreases during the interecdysial period. The incorporation of ^{14}C amino acids into proteins followed a similar pattern, but the increase in protein synthesis occurred approximately a day later than the increase in RNA synthesis. A high level of protein synthesis is reached at the onset of spinning, but it then declines during the pharate pupal stage.

It is crucial that the composition of incubation medium be controlled rigidly; e.g., the Ringer:blood medium gave a much lower rate of isotope incorporation and efforts to increase the incorporation rate

by increasing the isotope concentration 20 fold enhanced the rate only a third as much as with Ringer
alone. This may be due to larger pool of unlabeled
amino acids in the blood competing with isotopically
labeled amino acids. Also, the blood may contain an
inhibitory factor, or else the ionic milieu of insect
Ringer as such, is more conducive to RNA and protein
syntheses (Patel and Kroeger, 1972). Except for the
above experiment, all incubations were carried out in
insect Ringer.

The effects of DME, ecdysone and their mixtures
were examined at varying concentrations. The results recorded in Table 1 permit the following conclusions.

RNA synthesis:

1. Both DME and ecdysone increase the capacity
of discs to synthesize RNA.

2. The degree of stimulation depends on the
stage of the development of larva, e.g., stimulation
due to ecdysone was greater when the insects were
injected during the middle of the instar (3 day old)
as compared to the pre-ecdysial period (5 day old).
However, the discs are less sensitive to DME in 3-
day-old larvae as compared to the 5-day-old larvae.
Apparently, discs are more sensitive to injection of
hormone when the endogenous titer of that hormone is
low.

3. The mixture of the two hormones, however,
showed considerably less or no stimulation.

4. Very high doses of ecdysone on day 3 reduced
the amount of RNA synthesis.

Protein synthesis:

5. The capacity for protein synthesis did not
increase within 12 hrs. following ecdysone injection.

6. Protein synthesis was stimulated by DME and
it is stage-specific. The response was greater when
larvae approached ecdysis.

7. High doses of DME had no stimulatory effect
in 3-day-old animals.

Essentially the same results were obtained when
DME and ecdysone were substituted with more polar
LJM and ecdysterone.

The observation that more RNA and protein synthesis could be induced by either JH or ecdysone and
that their mixture fails to achieve expected stimulatory effect poses an interesting dilemma. One could
try to rationalize the observed data to fit one or
many concepts that could be tested. We have tried
to view this phenomenon in the light of the central
dogma that DNA transcribes RNA which in turn is used
for the translation of proteins. Hypothetically, we
have visualized that during normal growth these
processes are in a suppressed state. Growth is
maintained by two hormones which control two specific
but different sites, e.g., transcriptional and translational. These regulatory sites could be mutually
limiting, i.e., one is always inoperative while
another enhances the synthetic machinery. Thus,
application of one hormone when it is endogenously
absent or low, activates the site and stimulates
protein synthesis. However, if two hormones are
applied simultaneously, both sites become operative
and a suppressed state is attained. We may ask if
these sites are the same and inquire as to where
they would operate in the hierarchy of DNA-RNA-pro-
tein synthesis. The following preliminary experiment
was carried out to explore this phenomenon. Inhibition of macromolecules was affected by use of mitomycin for DNA, actinomycin D for DNA-dependent RNA
and puromycin for protein synthesis (Krishnakumaran
et al., 1967). Subsequently, hormones were administered and synthetic capacity of the wing disc examined.

We used 5.0, 1.25 and 200 µg of mitomycin C,
actinomycin and puromycin respectively, per g live
weight of the IV instar 5-day-old caterpillars. These
dosages are nontoxic but inhibitory. In this pre-

liminary experiment no attempts were made to attain
the same degree of inhibition. Twelve hours after
injection the larvae were injected with either 3.0
µg ecdysone, 100 µg DME or a mixture of two as described earlier. Then after 12 hrs., the animals
were dissected and wing discs incubated with RNA
and protein precursors and processed as in earlier
experiments. The results (Table 2) indicate that
ecdysone stimulates RNA synthesis in mitomycin treated cells, but it is incapable of reverting the inhibition of DNA primed RNA and protein syntheses.
Thus, ecdysone may act on the DNA synthetic apparatus (Madhavan and Schneiderman, 1968). However,
whether it acts by complexing with protein-SH groups
in an allosteric fashion, by binding to the histones
to increase the competence for gene expression, or
simply provides a proper and selective ionic or H+
environment for DNA-strands, remain speculative
possibilities, which are being explored in my laboratory. DME does not stimulate RNA synthesis in mitomycin inhibited cells. DME also prevents ecdysone
from acting to counter-inhibit mitomycin inhibition.
Whether this is accomplished by interacting with the
mitomycin sites or directly with ecdysone remains
to be investigated.

Protein synthesis was inhibited maximally by
puromycin, followed by actinomycin and mitomycin.
The ability to reverse this inhibition by the
hormones varied. DME was least effective in restor-
ing activity while a mixture of the hormones was
most effective. The mitomycin induced inhibition was
counteracted most by ecdysone. In fact, slight RNA
synthesis and considerable protein synthesis were
stimulated by ecdysone in mitomycin treated discs.
Irrespective of the antibiotic employed for both RNA
and protein syntheses, the mixture of the hormones
showed an averaging response, quite unlike that of
the regulatory response observed in the studies without antibiotics.

THE DEVELOPMENT AND DIFFERENTIATION

At the termination of the growth period, the holometabolous insect enters a phase of tremendously dynamic changes. Its behavioral pattern alters, the feeding ritual halts, the insect seeks a cloistral site and initiates spinning. In the protective silky enclave, it shortens its elongated body to an oval form. The covert infolded epidermal imaginal structures of the larvae are everted, the larval skin is then shed and replaced by the pupal cuticle. Thus, we witness the overt metamorphic drama, although it is not yet complete, as many internal organs undergo further differentiation. Tissues and cells undertake new functional responsibilities and their chemical and macromolecular constituents change. Among them are the proteins, which convey the gene expressed specificity and carry out specialized metabolic processes required during development and differentiation. With the notion that investigation of the proteins would reflect changes of these phenomena, we explored the careers of individual proteins, their specific activities, rates of syntheses, and when and where they are made and transported.

Materials and Methods

The saturniid silkworm _Hyalophora_ _cecropia_ was the experimental animal for this investigation. The developmental stages investigated included: eggs, V instar larvae, pupae before and after diapause, pharate and emerged adults.

The details of the collection of samples, their preparation for protein determination, acrylamide gel electrophoretic separation, densitometric evaluation of individual protein bands, and their syntheses by isotope labeling, etc., employed in this study are described in earlier publications (Patel and Schneiderman, 1969; Patel, 1971).

Though closer to reality, the _in vivo_ investigation of developmental processes in tissues and

organisms are more complex to interpret than _in vitro_
studies. Both have many merits, but _in vivo_ research
has many technical and methodological limitations.
These are discussed at length in the above papers.
I shall list a few which are pertinent for the present discussion.
- A protein band does not necessarily mean a single protein - it may represent many.
- Unless a protein is synthesized _de novo_, the
  determination of specific activity by isotope
  precursor incorporation could be masked by the
  concentration of that protein present at the onset of the experiment.
- The career of a specific protein in developmental studies should be considered in the light of
  its release, uptake and degradation in the host
  tissue and the organism.
- Concentrations determined by densitometry of the
  stained proteins reflect the dye binding capacity of a particular protein which may vary from
  one protein to another.

Keeping these limitations in mind, we shall examine: (I) the changes in the specific proteins during P-A development when differentiative processes
are expressed dynamically, and (II) regulatory influence of the insect hormones and their synthesis and
metabolism.

<u>Results</u>

(I) <u>Protein Synthesis During Development</u>

The investigation of the career of the proteins
includes blood as transport vehicle, fat body as a
storage depot, and epidermis and wings as tissues engaged in the differentiative changes. The changes in
the total protein reported in Fig. 3 show that blood
has the highest concentration followed by fat body,
wing and epidermis. However, the amount of proteins
synthesized in these tissues reveal that epidermis
was the most active tissue, followed by wing and

antenna (Fig. 4A), all ectodermal tissues, ovaries, blood and fat body (Fig. 4B) Note the reciprocal changes in both the total (Fig. 3) and newly synthesized (Fig. 4B) proteins in the blood and the fat body. How these observations reflect the career of specific proteins was further investigated.

<u>Blood</u>: (Fig. 5A, 5B)

Depending on the stage of development, the number of protein bands varied from 8 to 16, with 2 prominent bands, with R_F's of 32 and 40. Two slow-moving bands, with R_F's of 2 and 5, absent in V instar and early pharate pupae, appear between days 7 and 14 of adult transformation. A band, (R_F 12), seen in the larvae apparently shows a gradual shift into R_F 8 and increases in amount. Two bands, R_Fs 20 and 25, are present in appreciable amounts in stages L_8, PP_3-PP_6, and PA_5; they decrease in concentration between 7 and 10 days and are undetectable thereafter. The band R_F 32 is present in all stages. The band R_F 40 is found in highest concentration in larvae; it decreases and is not seen after day 18 of P-A transformation. There are a number of minor bands with R_Fs greater than 50 (52, 60, 78, 88 and 90). They are present in larval stages but progressively decrease in number and concentration.

This dynamic increase or decrease in the specific protein concentration raises questions, whether they are due to new synthesis or to utilization. This was examined by following the incorporation of labeled amino acids (Fig. 5B). It is quite clear that two proteins R_F 20 and 40 which disappear at a faster rate during development are not synthesized, but are observed at the top of the gel during the adult stage. Of particular interest is the protein band R_F 32, the female protein, which is synthesized continuously and at a higher rate than others.

The changes observed in the blood should reflect the performance of the tissues as the P-A is a non-feeding state and the amount of protein synthesized

by the blood cells (unpublished observations) could
not account for the observed magnitude of change.
Hence, the career of proteins were examined in
several tissues and their rates of synthesis, utilization or release into blood were studied during P-A
transformation.

<u>Fat body</u>:   (Fig. 6A, 6B)

Depending on the stage of development, up to 16
bands are displayed by the fat body. Of these, 4
strong bands are not detected before diapause or at
the pre-adult emergence period, exhibiting a pre-
ponderance only during the pharate stage. One of
them, R_F 25, occurs in high concentrations in diapause and throughout P-A development. Among other
tissues examined, only epidermis exhibits this band.
With development, this and band R_F 20 shift in migra-
tion and eventually they fuse on day 18 of transformation. This phenomenon was observed in 8 animals
examined. Unlike major blood protein bands, R_Fs
32 and 40 are present only in the pharate adult. The
general pattern of minor bands with R_F greater than
50 is different from that of blood, although one
band, R_F 68, is present in both tissues. In contrast
to the rest of the tissues, the adult fat body has
only 1 or 2 distinct bands, indicating the disappearance of most of the proteins. Occasionally 2
bands, R_Fs 3 and 5, are detected at the termination
of spinning and at onset of pupation. They are apparently short-lived and not detected in any other
tissue.
Very little, if any, synthesis occurs in the
pharate adult fat body (Fig. 6B). Indeed, the fat
body is depleted (Telfer, 1965) and the bulk of its
proteins used up before adult emergence, hence suggesting its function as a storage depot to supply the
precursors to other tissues. How precisely this
occurs is not understood. For instance, band R_F 40
is not labeled either in the blood or in the fat body
of the adult, hence the fat body could release this

protein into blood, except that it was present in the blood prior to its appearance in the fat body (compare Figs. 5A and 6A). This suggests that protein R_F 40 was transported to the fat body from blood. If this is the case, then band R_F 32 also should behave similarly. However, it is found labeled in the blood but not in the fat body. This is difficult to explain unless there is another protein in blood with the same R_F which is taken up by the fat body. In any case, the fat body has to be highly selective to recognize specific proteins. Whatever the mode of action may be, it remains quite puzzling that the proteins were taken up initially if the fat body has to return these proteins at some later time to other tissues, via blood. Only in part could one visualize the fat body's selective uptake of the non-labeled proteins (R_F 32 and 40). It could convert them, without incorporating labeled amino acids, into other proteins (R_F 20 and 25) or the latter ones could be degraded into the former ones and retained. This new configuration could render them storageable and suitable for specific degradation. Possibly, the degraded proteins are modified in such a manner that they utilize labeled amino acids to form derived proteins of the blood and other tissues as reflected in their reciprocal protein contents (Fig. 3 and 4B).

<u>Wings</u>:   (Fig. 7)

In contrast to blood and fat body, every wing protein incorporates labeled precursor. The wings and epidermis are not only engaged in multiferous activities, but they also show the earliest synthesis in a temporal sense. Synthesis in the wing is minimal by day 18 when wings are fully formed, while synthesis in the epidermis, blood, fat body and ovaries (to be discussed later) continues beyond day 18, suggesting a correlation of morphological and biochemical events during development. It may be noted that bands R_F 32 and 40 are labeled in wings and it is likely that one of them, R_F 32, could be released in-

to the blood. If this is the case, then R_F 40 must
be retained selectively. Most of the minor proteins
are found to be short-lived. Wings and other ectodermal tissues offer very exciting research possibilities.

(II) <u>Hormones and Protein Synthesis</u>

The temporal changes in synthesis and transport
of proteins during development provide a background
with which patterns altered by hormones or other
factors can be compared. We have employed several
different approaches and experimental maneuvers so
that we can better understand hormone action and
create a realistic concept. I shall discuss the
following specific areas: (A) JH and ecdysone assays
and their molecular significance, (B) perfusion techniques and hormones and other agents, (C) hormones
and reproduction, (D) fate of labeled hormones.

(A) <u>JH ecdysone assays and their molecular sig-
nificance</u>:

The most commonly used JH assay involves treat-
ing freshly molted <u>Tenebrio molitor</u> pupae with candidate material and observing the retention of pupal
characteristics over a period when control insects
become adults. The probability of attaining 100%
retention of pupal characteristics or perfect second
pupae depends on the type and concentration of the
compound as well as precise timings of application.
The 2-4 hr. old pupae are most sensitive. This sensitivity is lost after about 48 hrs., suggesting that
with age, the cells are committed to differentiation
and cannot be affected by JH. Also, there seems to
be a good correlation between the JH activity periods
and low ecdysone titer (Shaaya and Patel, in preparation). The analysis of protein synthesis in 2 and 24
hr. pupae treated with DME and ecdysone is shown in
Fig. 8 and allows us to draw the following conclusions.

1.   DME inhibits all ongoing protein synthesis.
2.   Greater inhibition is obtained in freshly molted pupae than older ones.
3.   Reduced inhibition is observed in a specific protein band (at the top of the gel) in older pupae.
4.   Ecdysone exerts a less drastic influence than JH, and is slightly inhibitory in younger but stimulatory in older pupae.
5.   Except for a slightly inhibited band at the top of the gel and R_F 70, ecdysone does not exhibit greater specificity.
6.   Not reported in the figure, ecdysone treated pupae became adults 10-25% sooner than controls, while no DME treated pupae emerged as adults.

From the data it seems that both hormones affect protein synthesis in a more generalized fashion and lack definite and pronounced specificity in short term experiments.

In reality, JH retains the morphological characteristics in a subtle way. Thus, a perfect second pupa is not a dead pupa; on the contrary, it has made a cuticle in a modified way. The epidermal cells secrete an 'outer cuticle' under which is formed another 'inner cuticle' when treated with JH. The amino acid composition of these cuticles are compared with others occurring during normal metamorphosis (Table 3). Of particular interest is the comparison of pupal with inner cuticle to P-A with outer cuticle. The relative percentage compositions of these pairs show greater similarities than comparisions of pupal with P-A and inner with outer cuticles. The outstanding effect of JH is expressed by the incorporation of previously unobserved cysteine (cystine) into cuticle protein, and lower methionine incorporation in the outer cuticle as compared to P-A cuticle. It is not known if these effects are due to misreading the protein coding mechanism or enzymatic interference in sulfur crosslinkages required for structural configuration of cuticular proteins. Serine, threonine, aspartic acid, tyrosine and isoleucine exhibit greater discrepancies, while leucine shows a constant

level in all cases. The precise molecular significance of the incorporation of these specific amino
acids is not clear at present. However, selection
of a certain pair, e.g., tyrosine: leucine, could
provide a sensitive measure of hormonal action of
protein synthesis as employed by Ilan, et al. (1970).
They reported that gene expression is mediated by JH
at the translational level, that it involves the
appearance of new tRNA, that in part, the in vitro
translation of cuticular proteins is tRNA dose
dependent and that the decrease in tyrosine: leucine
ratio of synthesized proteins could be assigned to
JH action.

   We have continued the exploration of this phenomenon in search of the site or sites for the specificity in the translational regulation (Patel, Ilan
and Ilan, in preparation). Freshly molted Tenebrio
pupae were allowed to age for 12 and 72 hrs. and
then were treated with either DME or ecdysterone.
Ribosomes were prepared from microsomes by sodium
deoxycholate treatment and the proteins stripped off
the ribosomes with 4M LiCl and 8M urea. The ribosomal proteins in 8M urea were separated on 4.5%
acrylamide gel electrophoresis. The densitometric
tracings were shown in Fig. 9. The following observations are made.

-   During 12-72 hr. normal development the riboso-
    mal protein pattern remained unchanged, al-
    though the total amount of proteins increased
    somewhat in older animals.
-   In DME treated younger animals one band (No.6)
    was missing, while another (No.8) was lost in
    the older group. Also, several smaller pro-
    teins (Nos. 14-17) were depleted in both age
    groups.
-   Ecdysterone treatment decreased band No. 3, 5,
    9 and 10 in younger pupae. The ribosomal pro-
    teins from 72 hr. insects failed to provide
    distinct bands in 5 separate experiments. It
    was difficult to decide whether this was a gen-
    uine effect of ecdysterone or artifact.

The insect ribosomal protein patterns observed by us are quite similar to $\underline{E}$. $\underline{coli}$ (Nomura $\underline{et}$ $\underline{al}$., 1970; Kurland $\underline{et}$ $\underline{al}$., 1969). These authors have assigned values of translational function to certain proteins and others to be responsible for maintaining structural integrity. DME and ecdysterone-affected proteins fall into two sets, each one different and each including both functional and structural proteins. We have also observed that the ^{14}C-leucine incorporation pattern into ribosomal proteins of DME treated insects is reciprocal to those treated with ecdysterone. How this correlates to the above concept is unclear. Whether specific hormonal binding could be responsible for the expression remains to be seen and is worth investigating.

Routine ecdysone activity determinations (Shaaya and Karlson, 1965) are done by ligating a prepupal stage at a critical time such that the portion posterior to the ligation does not receive the signal to pupate from the anterior endocrine organs. If either the ecdysone or an active candidate material is administered to the posterior portion, pupation ensues. We (Patel and Sehnal, 1972) carried out similar experiments with the wax moth $\underline{Galleria}$ $\underline{mellonella}$ and investigated the effects of ecdysone on the protein synthesis under these conditions and the results are reported in Fig. 10A and 10B. The data revealed that:

1. Amount of total protein fluctuated only slightly in the ligated-nondifferentiating-isolated abdomens, over a period of 25 days.

2. Ecdysone evoked a slight, if any, response in the total protein within 2 hrs. after injection.

3. Protein synthesis continues to take place in these abdomens as labeled amino acids are incorporated.

4. The rate of incorporation increases 3.5 to 4 fold due to ecdysone treatment, indicating definite stimulation of protein synthesis.

5. Increased incorporation due to ecdysone was observed in all the proteins separated by gel electrophoresis, indicating lack of specificity of its

action under these conditions, i.e., lack of brain
and other anterior endocrine organs.

These studies point to the overall effects on
protein synthesis by JH as inhibitory and that of
ecdysone to be stimulatory. However, their actions
lack the specificity to turn off or on individual
proteins normally expected for selective gene expression. I shall later discuss the relative merits
of these observations.

(B) <u>Perfusion Studies</u>:

During metamorphosis qualitative and quantitative changes occur in blood proteins (Patel and
Schneiderman, 1969). We observed that (i) some proteins were present in all three stages, i.e., larval,
pupal and adults, (ii) some were stage-specific and
(iii) most, but not all, were in a dynamic state of
synthesis and transport.

The questions we asked were: 1) If we deplete
the blood proteins by perfusion (Patel and Schneiderman, 1969) would the low concentration of proteins
stimulate synthesis and release of proteins? 2) What
course of action would occur to reconstitute the
blood? 3) What influence do the regulatory molecules
like hormones and other factors exert?

The same techniques used by Patel and Schneiderman (1969) and Patel (1971) were employed, and the
data obtained from 4.5% gel electrophoretic analysis
provided the following information. (For identifying
specific protein bands refer to Table 4.)

1. The pre-perfusion diapausing pupae are cap-
able of synthesizing proteins. Unchilled, unchilled
debrained and previously chilled debrained pupae
showed that (i) under no circumstances were the 3
fast moving larval specific proteins observed or
synthesized, (ii) of the six major proteins, unchilled pupae showed synthesis of all but two, including
the carotenoid protein, (iii) debraining stimulated a
higher rate of synthesis in most bands, including the
female protein. Observations over an extended period

showed that a slow moving adult protein was not synthesized while others were synthesized at a lower rate (Patel and Schneiderman, 1969).

2. The post-perfusion synthesis falls into three categories: (i) most every major protein (R_F 16, 25, 30, 38, and 46) normally found in pupal blood remain unstimulated; (ii) synthesis of adult protein (R_F 2) was stimulated; (iii) and three minor larval proteins (R_F 56, 63 and 77) normally absent, appeared in blood and their synthesis showing the greatest degree of stimulation.

3. Synthesis affected by various regulatory maneuvers was multiferous, and provided the following information: (i) perfusion stimulated synthesis up to 20-fold, (ii) synthesis was lowest in non-perfused debrained animals, (iii) non-perfused sham operated pupae behave like normal developing animals, (iv) DME prolonged and ecdysone enhanced the disappearance of proteins from blood, and similar temporal effect on the protein synthesis was also observed. This type of regulation occurred in both perfused and non-perfused animals, (v) the female protein (R_F 38) was the most stimulated protein resulting from perfusion and hormonal influence. It responded to JH stimulation during an earlier phase of development and to ecdysone later, although to a lesser degree (Fig. 11).

4. Effect on specific protein synthesis: (i) the carotenoid protein, band 3 (Table 4) was most always present in animals with brains but absent from the debrained ones. Implantation of additional brains from diapausing pupae, however, did not reinstitute this protein, (ii) compounds with JH activity, e.g., DME, FME, and _Cecropia_ oil, depleted carotenoid protein and concomitantly induced a new protein (R_F 25-30). Independently, the latter observation has been made in _Oncopeltus_ by Bassi and Feir (1971), (iii) the carotenoid protein was found in insects receiving ecdysone, K^+, Mg^{++}, insect Ringer, 10% EtOH and _Cecropia_ oil from allatectomized animals. On the basis of these observations, a broad generali-

zation may be made, that ecdysone experts a positive
and JH a negative control on the carotenoid protein,
while the brain oversees its regulation. The loss
of carotenoid protein and appearance of a new one
suggest that either a totally new synthesis or, more
likely, a substitution or removal of a prosthetic
group is regulated by JH. The entire regulation may
be mimicked by more than one molecule, although high
specificity may be executed by few hormonal compounds.

(C)   <u>Hormones and Reproduction</u>:

It was observed (Patel, 1971) that the major
vitellogenin (R_F 38) was synthesized during P-A
transformation and was transported to the follicles
(Anderson and Telfer, 1970; Anderson, 1971). The
effect of JH and MH was studied by injecting BGE,
inokosterone or a mixture, and labeled amino acid
precursors at the time of P-A initiation. The females were sacrificed on the second day after adult
emergence, the ovaries dissected out, and the proteins analyzed. The results in Fig. 12 show that all
the treatments increased the concentration of labeled
protein in the follicles. The effect of the mixture
was particularly surprising in view of its autoinhib-
itory effect reported earlier. The total amount of
protein synthesized was 38,700, 52,500, 46,400 and
68,600 dpm/mg protein for the control, BGE, inokosterone and their mixture, respectively. Similar
effects were also observed for the blood female protein band R_F 38. This observation could not be supported entirely from the experimental data at hand.
Although it seems that ecdysone favors early formation of yolkless ovaries (Williams, 1968), we have
shown that JH facilitates vitellogenin synthesis.
This agrees well with the temporal make-up of endogenous titers of ecdysone (Shaaya and Patel, in preparation) and of JH (Gilbert and Schneiderman, 1961).
Hence, in concert they could promote greater synthesis, as their individual actions are upon different

target cells at specific times.

It seems quite clear from the above observations that hormones enhance synthesis and may, although not experimentally demonstrated, also promote the uptake. In view of the complex interrelationships of the role of the corpora allata, JH, egg maturation, synthesis and uptake of vitellogenins involved (to be discussed later) we decided to clarify the role of JH in the uptake of the vitellogenins, distinct from synthesis. Vitellogenins were prepared by injecting chilled diapausing pupae with labeled amino acids as described earlier. The ovaries were dissected out on day 17 of P-A development, homogenized to extract soluble proteins, dialyzed exhaustively to remove precursors, lyophilized and reconstituted to give 100 mg protein/ml insect Ringer. _In vitro_ incubations were carried out at 29°C in a gyroshaker. Three ovarioles from 17-day pharate adult were incubated for 4 hrs. in a 3 ml protein solution with or without 5 µg of CJH/ml. At the end of the incubation, the ovarioles were gently but thoroughly rinsed, and a section consisting of 3 follicles each was removed and the protein uptake determined. The results showed that the radioactivity was about twice as great in JH treated follicles as in the controls.

To eliminate the possibility that the preparation of vitellogenins used in the incubation above may contain several proteins synthesized by either nurse cells or follicular cells which could have been reused by those cells and may not represent a true vitellogenin uptake, we carried out further experiments. The labeled proteins were further separated by free flow electrophoresis, dialyzed, lyophilized and the identity of the major vitellogenin reconfirmed by gel electrophoresis. The purified protein then was used for _in vitro_ incubation as stated earlier. Aliquots of 10 µl samples were removed from the incubation medium, during and at the end of the experiment, at which time the follicles were removed, rinsed and the protein quantitatively extracted. Two hundred µg of protein was applied per gel for electro-

phoresis. The separated bands were stained, cut into
18 equal parts and counted. The results are shown in
Fig. 13. The results clearly demonstrate that JH
influences the uptake of the major protein, as it is
depleted from the surrounding milieu.

(D) <u>Fate of Labeled Hormone</u>:

A question of considerable significance that
always holds a center position is how do hormones
regulate and at what sites? I shall attempt to
answer the last question. Again, I shall limit my
discussion to the pertinent observations and set
aside the details which are underway to be published
elsewhere.

The synthetic CJH was tritiated by catalytic
reduction with tritium gas (New England Nuclear Co.,
Boston, Mass.), and was extensively purified so as to
contain only ^{3}H-CJH. This purification was necessary
because catalytic labeling produces side reactions
and by-products. On the basis of preliminary action-
time curves, an hour of incubation time was considered satisfactory. The ovaries with attached fat body
from day 17-18 pharate adult female were dissected
out in insect Ringer and incubated in 0.3 ml of
medium containing blood:Ringer (1:1) and 10^6 dpm of
^{3}H-CJH/ml. The incubation was terminated after an
hour by fixing in a cold osmium tetroxide insect
Ringer solution. Due to the lipoidal nature of JH,
there were considerable difficulties in retaining the
label within the tissue. However, suitable steps
were worked out. The final embedding was done in
Epon and 3 µ sections were cut for light microscopic
and thin (600 $\overset{\circ}{A}$) sections for electron microscopic
examination. The following observations were made
from the autoradiographs:
1. The fat body consists of lipid spheres sur-
rounded by protein droplets. There was an intense
labeling of the lipid spheres and considerably less
of the protein droplets. Occasionally the lipid
spheres were found to be hyaline rather than greyish

in filled regions. Under these conditions, only the
osmium-stained grey region had the grains.

2. The distribution in the follicular cells
was restricted to the cytoplasm. There were 10% or
less grains in the nucleus and they were not found on
the chromatin material.

3. No grains above background were observed in
the intercellular spaces.

4. There was little prepondrance of grains on
the pinocytotic membrane.

5. The grains inside the follicle were extensively located in the yolk spheres and less than 5%
were found either in the lipid droplets, which sur-
round the protein spheres, or in the background
matrix. This was not the case in milkweed bug oocytes (unpublished data). It has more osmium-stained
lipid droplets than Cecropia. The autoradiographs of
the earlier phase of oocyte growth show extensive
labeling of lipid droplets, while advanced stages exhibit more radioactivity in the protein spheres.

6. The intensity of the grains decreased to-
wards the center of the follicle, where the distribution of grains assumed a ring format, probably indicating a frozen event in the entry of the labeled
molecule.

DISCUSSION

The biology of the insect occurs in one endless
continuum, subdivided into two phases of growth and
metamorphosis; the latter characterized by three
discernible stages, egg, larva and adult, with a
transitory pupal stage following larval in some insects. The basis for the entire hierarchy lies in
the genome of the egg. In multicellular animals the
expression of genome has to be temporal to satisfy
specific functionality at a precise developmental
stage. Such a control is achieved by regulatory
molecules secreted by special endocrine glands. There
is overwhelming evidence that in insects, it is monitored by two hormones, juvenile hormone and ecdysone

(Novak, 1965; Wigglesworth, 1970). It seems that JH regulates the growth and ecdysone provides the conditions for growth. JH attains its function by availing the cells the opportunity to assimilate components from its surroundings by applying subtle brakes on the processes which otherwise would differentiate the cells to assume newer responsibilities, whereas ecdysone, in turn, interrupts the growth to initiate, promote and execute the genomic message. Both hormones try to utilize the energy resources of the cells without wasteful duplication. It is this interplay of two hormones, how they function in concert and out of synchrony to regulate the molecule events, which are expressed as basic biological phenomena of growth, development, differentiation, and reproduction, in which we are interested.

The present study provides five kinds of evidences that JH and ecdysone regulate the growth during larval stages. The first observation is conceptually diagramed in Fig. 14 and shows that changes in RNA and protein synthesis are cyclical and most likely could be coupled with the changes in JH and ecdysone. The second observation (Fig. 7) reinforces the above evidence, and shows that the levels of RNA and protein syntheses increase progressively with the age, and that a decreased JH titer and an increased ecdysone titer are required for molting. That is, both hormones do not occur in appreciable amounts simultaneously during the growth phase. The third line of evidence is that of the tremendous changes in the rates of RNA and protein syntheses which are attained by exogenous administration of these hormones or analogs, and that the amount of response depends on the temporal stage of the larvae between ecdysial events. Thus, the greatest effect was seen when the concentrations of endogenous hormones were the lowest. Injection of 3 µg/g ecdysone into a 3-day-old IV instar larva, which apparently has little ecdysone, increased RNA synthesis about 10-fold over the controls, whereas a similar injection into IV instar larvae just prior to ecdysis, which has a high titer of ecdysone, increased RNA synthesis only 2.5-fold. The fourth rather interesting observation shows that the stimulatory effect of either of the hormones is nullified when both are applied together. The fifth item of evidence demonstrates that the action of the antibiotic mitomycin C, which leads to the scission of DNA strands and inhibits DNA synthesis, could be counteracted by ecdysone but not by DME, and ecdysone becomes ineffective in the presence of DME. This reversal is measured in the synthetic capacity of the wing discs for making both RNA and proteins. Ecdysone cannot revert inhibition induced by actinomycin D or puromycin, hence it must be specific for DNA synthesis. DME inhibition is facilitated in the discs subjected to these antibiotics. The mixture averages the effect of the individual hormones for macromolecular synthesis in discs pretreated with any of the three antibiotics. This observation could be interpreted as a mere physical and additive response in contrast to the regulatory control implemented by the mixtures in the discs without antibiotic treatment. Thus, the latter phenomenon seems to be truly regulatory and is endowed with specificity for macromolecular synthesis. It may be stressed here, that data obtained from any experimental manipulation provide valuable information but represent the naturally occurring phenomenon only in part, and then out of context. The chemical composition of second pupal cuticle (JH effect) and non-specific stimulation of protein synthesis in pupae (ecdysone effect) in ligated larvae, are good examples.

Let us examine these observations and lines of evidence in a broader perspective of insect biology. The insect life cycle has two chief functional stages, immature larval stage to grow and mature adult stage to reproduce. The overt growth is essentially a continuous process massing the nutrients necessary for the maintenance of life-supporting processes and in preparation for the reproduction. The heterogenic nature of growth in organs and tissues of larval

stages precludes qualitative differentiation, and
with few exceptions it is attained by an increase in
the size of cells, not the number. Continuous growth
must be interrupted by ecdysial events to facilitate
accommodation of the increasing mass enclosed in the
semi-rigid protective cuticle. This is attained by
the interactions between cells and exogenous factors
and regulators, such as hormones. The programmed
growth rate from the egg to the last larval stage is
a result of a progressive decline of the JH titer.
This negative correlation suggests that the precise
mechanism may involve the release of a suppressed
cellular and molecular state continuously maintained
by JH. If this is true, then increased protein synthesis during growth, together with JH, may serve to
regulate the macromolecular synthesis by feedback
inhibition. This may involve all regulatory processes, but chiefly the translational controls, as protein synthesis is affected directly. The reciprocal
increase in the amount of ecdysone for its desired
effectiveness for ecdysis is required to compensate
for the increased biomass due to growth. The role of
ecdysone is more specific and dramatic. It has to
evoke a digressed response in the steady pattern of
JH action. Ecdysone effectively accomplishes this
by acting at the sites of DNA synthesis, and hence
must involve predominantly transcriptional mechanisms.

Viewing growth at the molecular level, we are
witness to the interaction of hormones with essentially non-differentiated cells. These cells are
not committed to take on new responsibilities, although some of them seem to assume such function at
ecdysis. Once committed, they reach a point of no
return, and thus, pupal cells cannot be reverted to
the larval type, in totality.

Development

The concept of development is inclusive of differentiation. The study of insect development shows

the dynamic unfolding of events of L-P-A transformation when cells differentiate to assume newer and
more complex functions. These phenomena are viewed
as changes in soluble proteins, but could also be
studied for any other molecule.

Prior to the onset of spinning, larvae take
their last meal and provide us with a closed system.
The blood and the fat body protein changes under this
condition are of particular interest. Their total
proteins show reciprocal patterns that ensue with
pupation and lasts till the end of pharate adult life
(Fig. 3). These tissues contain the highest concentration of proteins but demonstrated the lowest rates
of synthesis (Fig. 4A-B), suggesting that they are
essentially protein depots, blood being a mobile one.
In contrast, ectodermal tissues like epidermis and
wings are sites of active protein synthesis as they
engage in differentiative processes.

The changes in specific blood proteins are less
pronounced during L-P transition, which shows dramatic morphological alterations, than during the P-A
where little overt change occurs. Of the 7 major
proteins (the numbers and R_Fs are in Table 4), there
was no change in 5 during L-P transformation. Two
bands (Nos. 4, 5) were drastically reduced between
5-7 days and others (Nos. 5, 6) were gradually depleted. Specific fat body proteins remain stored
during the major portion of the P-A stage. Of the 4
major proteins (Fig. 6) none were found in L-P or
pre-adult stages. The situation with the wings is
considerably different. It contains (Fig. 7) very
little of the major proteins, and most every protein
found is actively synthesized. It is practically
impossible to discuss here the role of minor proteins
individually. These observations demonstrate the
discrete role of each tissue in maintaining its own
integrity but staying in concert with its surroundings. Thus, it seems that greater specificity re-
sides within the cell rather than with proteins.

The generalized effect of JH and ecdysone in
Cecropia P-A development can be summarized as

follows: JH apparently acts on the cells at the time of initiation of adult development and stretches the growth, more so at the terminal end, and it prolongs the protein synthesis period; ecdysone, on the contrary, enhances the development, primarily by shortening the earlier events as also observed by Williams (1968) and as reflected in overall protein synthesis (Patel, unpublished data).

The specific effects of JH and ecdysone at the molecular level have been considered by many scientists (Karlson and Sekeris, 1966; Kroeger, 1968; Ilan et al., 1970; Williams and Kafatos, 1971; Ilan et al., 1972). Various concepts are provided on the basis of their experimental data and intelligent extrapolations, based on knowledge of gene expression in microbial systems. Which of these concepts would be operative in insects and to what extent will depend on crucial experiments to prove the claims made. One central theme, which is accepted by all, is that hormones act on the gene expression hierarchy. The disagreements could be in the refinements of the mode to induce specific and primary events, all the way from hormones changing the permeability of the cells to alter ionic contents, to loosening histone from the DNA proteins, to regulation of and binding with mRNA to specifically activating RNA polymerases, to affecting translational machinery that regulates the type of proteins made, etc. Also, attempts have been made to draw inferred correlations, i.e., showing occurrence of specific puffs during development and inducing these puffs by hormones and thus deducing gene activation (Clever, 1963; Kroeger, 1963; Laufer and Holt, 1970). In addition, attempts have been made to show how hormones reach the target sites (Chino et al., 1970).

A rather interesting observation needs to be made; in general, the published literature includes data on the effects of exogenous hormones without serious consideration of endogenous titers. If viewed thusly, during L-L transformation following ecdysis, we have a high titer of JH and low ecdysone and

the reverse before ecdysis. However, in <u>Cecropia</u>,
following L-P we have high ecdysone and no JH and
before P-A ecdysis, a high JH and low ecdysone. What
significance such milieu would play on the cell is
unknown at the moment.

One of the roles of JH is to promote egg maturation (Reviewed by Telfer, 1965; Wigglesworth, 1970),
as has been reported in various groups of insects
(Johansen, 1958; Engelmann, 1969; Bell and Barth,
1970; Pan and Wyatt, 1971). However, removal of
corpora allata in most <u>Lepidoptera</u> in the pupal
stage has not prevented the egg maturation. In fact,
removal of corpora allata from larvae in <u>Bombyx</u>
(Fukuda, 1944) did not prevent egg maturation, and
isolated <u>Cecropia</u> abdomen devoid of corpora allata,
but in the presence of prothoracic glands (Williams,
1961) did produce eggs. Other lepidopterous insects
are known to require the corpora allata for egg
maturation, as in the case of <u>Galleria</u> (Röller, 1962;
Sehnal, 1968), <u>Pieris</u> (Karlinsky, 1967) and <u>Danus</u>
(Pan and Wyatt, 1971).

The problem is too complex to be resolved by
one crucial experiment as it involves lepidopterous
species with varying biology in the adult insects.
Also, we have no information as to what the delayed
action of corpora allata secretions may be. In addition, if, when, and how much of an effect JH would
have on the nurse cells, follicular cells and pinocytotic uptake is important as they all contribute
to the final outcome. Information on any one of the
above may not explain the entire process, but would
open up the way to experimental testing. Our data
show that injection of JH, ecdysone, and their mixtures at physiological concentrations stimulate
vitellogenin synthesis and JH positively enhances the
uptake of the purified major vitellogenin <u>in vitro</u>.
Similar observations were also made for milkweed bug,
<u>Oncopeltus fasciatus</u> (unpublished). Not only does JH
stimulate these processes, but also the labeled JH is
picked up and grains in autoradiographs are preferentially located in fat body lipid spheres, follicular

cell cytoplasm and oocyte protein yolk spheres.
Selectively, more grains were found in milkweed bug
oocyte lipid droplets than <u>Cecropia</u>, indicating
species and group specificities.
   The following are the highlights of this presentation.

<u>SUMMARY</u>

- The growth phase of non-metamorphic stages requires a precise concerted regulation of both
   hormones in appropriate concentrations.
- Only at a low native hormone titer does the addition of individual hormone stimulate synthesis.
- Juvenile hormone and ecdysone regulate RNA and
   protein synthesis.
- Mixture of JH and ecdysone inhibits the expected
   stimulation.
- The metamorphic events coincide with profound
   changes in total and specific proteins - most
   likely linked to differentiation.
- There are stage and tissue specific proteins,
   which are synthesized at varied levels. Some
   are released in the blood and transported to
   other tissues.
- Perfusion offers an exciting possibility to attain
   highly stimulated environment for macromolecular
   synthesis to investigate role of hormones.
- Contribution of specific ions should be taken into
   account for the refined hormonal action.
- Brain (via corpora allata and prothoracic glands)
   exerts a suppressive directional control of protein synthesis.
- Brain seems to execute a power of discretion and
   oversees the specifications of JH and ecdysone
   actions.
- Specific reproductive processes are regulated by
   JH and ecdysone, and they may work in concert to
   satisfy the requirements of egg maturation.
- Both the pinocytotic uptake and synthesis of
   vitellogenins are affected by JH and ecdysone.

- For the expression of hormone action, the constituent chemical make-up of target cells is equally important to the presence or absence of hormones.
- Action of JH is repression-promoting, universal and drastic, often by feedback inhibition via increased protein synthesis.
- Ecdysone predominantly regulates DNA-function and directions of events.
- JH and ecdysone affect ribosomal proteins and thus infer translational regulation of gene expression.
- Amino acid analysis of cuticles from JH treatment show that the shed and newly laid cuticles are different, hence the old genes are not turned on _in toto_, but translation is modified considerably.
- _In vitro_ uptake of labeled JH indicates preferential sites where it may act.

ACKNOWLEDGEMENTS

This work represents, in part, the cumulative efforts of many associates and colleagues. My special thanks goes to Professor Howard Schneiderman, who introduced me to this fascinating field of biology and provided the opportunities for the investigations. I sincerely extend my appreciation to Dr. Kornath Madhavan, who collaborated in the wing disc studies, Drs. Joseph and Judith Ilan in molecular aspects, Dr. Edward Wat for his synthesis of the many compounds employed, and Mr. Richard Hebert for electron microscopic studies. Without the assistance of Bob Cowen, William Edds, Jr., David Thorpe and Cecelia North, this work would not have been accomplished. I also thank many colleagues who suggested and criticized on many aspects during our discussions and Dr. Edmund I. Stevenson for his valuable comments on the manuscript.

REFERENCES

Anderson, L. (1971). J. Cell Sci. 8, 735.

Anderson, L. and Telfer, W.H. (1970). J. Cellular Physiol. 76, 37.

Bassi, S.D. and Feir, D. (1971). Insect Biochem. 1, 433.

Bell, W.J. and Barth, R.H. (1970). J. Insect Physiol. 16, 2303.

Berkoff, C.E. (1969). Quart. Rev. 23, 372.

Bowers, W.S. (1969). Science, 164, 323.

Butenandt, A. and Karlson, P. (1954). Z. Naturforsch. 9b, 389.

Chino, H., Gilbert, L.I., Siddall, J.B. and Hafferl, W. (1970). J. Insect Physiol. 16, 2033.

Clever, U. (1963). Develop. Biol. 6, 73.

Clever, U. and Karlson, P. (1960). Exp. Cell Res. 20, 623.

Engelmann, F. (1969). Science 165, 407.

Fukuda, S. (1944). J. Fac. Sci, Univ. Tokyo Sect. IV, 6, 477.

Gilbert, L.I. (1969). Proc. 3rd Intern. Congr. Endocrinol. 340.

Gilbert, L.I. and Schneiderman, H.A. (1961). Gen. Comp. Endocrinol. 1, 453.

Ilan, J., Ilan, J. and Patel, N. (1970). J. Biol. Chem. 245, 1275.

Ilan, J., Ilan, J. and Patel, N. (1972). In "Insect Juvenile Hormones", (J. J. Menn and M. Beroza, eds.), pp. 43-68, Academic Press, New York.

Johansson, A.S. (1958). Nytt. Meg. Zool. 7, 1.

Karlinsky, A. (1967). C. r. hebd. seanc. Acad. Sci. Paris 264, 1735.

Karlson, P. and Sekeris, C.E. (1966). Acta Endocrinol. 53, 505.

Kopeć, S. (1917). Bull. In. Acad. Sci. Cracovie (B), pp. 57.60.

Kopeć, S. (1922). Biol. Bull. 42, 323.

Krishnakumaran, A., Berry, S.J., Oberlander, H. and Schneiderman, H.A. (1967). J. Insect Physiol. 13, 1.

Krishnakumaran, A.K. and Schneiderman, H.A. (1968).
    Nature 220, 601.
Kroeger, H. (1963). J. Cell. Comp. Physiol. 62, 45.
Kroeger, H. (1968). In "Metamorphosis", (W. Etkin
    and L. I. Gilbert, eds.), pp. 185-219, Appleton-
    Century-Crofts, New York.
Kroeger, H. and Lezzi, M. (1966). Ann. Rev. Entomol.
    11, 1.
Kurland, C.G., Voynow, S.J.S., Hardy, L., Randall,
    L. and Lutter, L. (1969). Cold Spring Harbor
    Symp. 34, 17.
Laufer, H. and Holt, T.K.H. (1970). J. Exp. Zool.
    173, 341.
Law, J.W., Yuan, C. and Williams, C.M. (1966). Proc.
    Nat. Acad. Sci. 55, 576.
Locke, M. (1964). In "Physiology of Insecta", (M.
    Rockstein, ed.), 3, pp. 380-466, Academic Press,
    New York.
Madhavan, K. and Schneiderman, H.A. (1968). J. Insect
    Physiol. 14, 777.
Nomura, M., Mizushima, S., Ozaki, M., Traub, P. and
    Lowry, C.V. (1969). Cold Spring Harbor Symp. 34,
    49.
Novak, V.J.A. (1965). "Insect Hormones", Metuen and
    Co., Ltd., London.
Pan, M.L. and Wyatt, G.R. (1971). Science 174, 503.
Patel, N. (1971). Insect Biochem. 1, 391.
Patel, N. and Kroeger, H. (1972). Insect Biochem.
    (in press).
Patel, N. and Madhavan, K. (1969). J. Insect Physiol.
    15, 2141.
Patel, N. and Schneiderman, H.A. (1969). J. Insect
    Physiol. 15, 643.
Patel, N. and Sehnal, F. (1972). (submitted for
    publication).
Röller, H. (1962). Naturwissenschaften. 49, 524.
Röller, H., Dahm, C.C., Sweely, C.C. and Troast, B.M.
    (1967). Angew. Chem. Intern. Ed. Engl. 6, 176.
Sehnal, F. (1968). J. Insect Physiol. 14, 73.
Shaaya, E. and Karlson, P. (1965). J. Insect Physiol.
    11, 65.

Schneiderman, H.A. (1967). In "Methods in Developmental Biology", (F. H. Wilt and N. K. Wessels, eds.), pp. 753-766, Crowell, New York.

Schneiderman, H.A. and Gilbert, L.I. (1957). Anat. Record 128, 618.

Schneiderman, H.A., Krishnakumaran, A., Kilkarni, V.G. and Friedman, L. (1965). J. Insect Physiol. 11, 1641.

Telfer, W.H. (1965). Ann. Rev. Entomol. 10, 161.

Wigglesworth, V.B. (1934). Nature 135, 725.

Wigglesworth, V.B. (1935). Nature 136, 338.

Wigglesworth, V.B. (1936). Quart. J. Microscop. Sci. 79, 91.

Wigglesworth, V.B. (1954). Quart. J. Microscop. Sci. 95, 115.

Wigglesworth, V.B. (1970). "Insect Hormones", H. W. Freeman and Company, San Francisco.

Williams, C.M. (1956). Nature 178, 212.

Williams, C.M. (1958). Sci. Am. 198, 67.

Williams, C.M. (1961). Biol. Bull. 121, 572.

Williams, C.M. (1968). Biol. Bull. 134, 344.

Williams, C.M. and Kafatos, F. (1971). Mitt. Schweiz. Entomol. Ges. 44, 151.

TABLE 1

Response of the Wing Discs of <u>Ricini</u> Silkworm to Insect Ringer,
Peanut Oil, Dodecyl Methyl Ether, Ecdysone, and Their Mixtures

| | Radioactivity (dis/min)/µg dry wt. | | | |
| | RNA Synthesis* | | Protein Synthesis** | |
Treatment	3 day[#] Dis/min±S.D.	5 day[#] Dis/Min±S.D.	3 day[#] Dis/min±S.D.	5 day[#] Dis/min±S.D.
Control – insect Ringer	13.85±10.64	22.89±12.33	50±30	180± 60
Control – peanut oil	13.32± 7.98	27.41±13.70	70±40	100± 30
Control – 10% EtOH	5.32± 5.32	8.22± 2.74	50±10	130± 60
DME, 10 µg/g[##]	79.80± 5.32	8.22± 5.48	150±40	470± 80
DME, 100 µg/g	93.10± 7.98	102.78± 6.85	330±60	660±130
DME, 1000 µg/g	101.38± 7.98	208.31±41.11	70±60	900±110
Ecdysone, 0.3 µg/g[##]	63.84±18.62	19.18± 8.22	60±10	40± 30
Ecdysone, 3.0 µg/g	266.00±26.60	65.78± 5.48	90±50	80± 20
Ecdysone, 30.0 µg/g	114.38±21.28	49.33±13.70	130±30	140± 20
DME + ecdysone, 10 µg + 0.3 µg/g[##]	21.28±10.64	41.11±10.66	40±20	90± 10
DME + ecdysone, 100 µg + 3.0 µg/g	13.32± 5.32	43.85±10.96	250±30	160± 30
DME + ecdysone, 1000 µg + 30.0 µg/g	37.24±10.64	5.48± 8.22	70±10	140± 10

The measurements represent the capacity of the 3- and 5-day-old IV instar wing discs to synthesize RNA and proteins 12 hr. following treatment. The values represent averages of 6 to 8 discs in each treatment.

*RNA synthesis computed on the basis that the incubation medium contained 10^5 dis/min.

**Protein synthesis computed on the basis that the incubation medium contained 10^4 dis/min.

#Age of wing discs in days of IV instar larvae.

##Active ingredient injected into larvae/g live **wt.**

TABLE 2

RNA and protein syntheses in the wing discs of 5-day-old, IV instar <u>Samia</u> <u>cynthia</u> <u>ricini</u> larvae subjected to mitomycin C, actinomycin D and puromycin and followed by the injections with ecdysone, DME and their mixtures. (See text for details.)

	dpm/µg disc tissue/1 x 10^6 dpm of precursor			
Treatment	Mitomycin C	Actinomycin D	Puromycin	Control
	RNA			
Ecdysone	3,080	342	342	2,530
DME	1,830	375	308	6,950
Mix	2,620	820	502	3,670
	Protein			
Ecdysone	40,400	3,600	1,320	3,123
DME	8,320	2,080	640	4,097
Mix	10,600	6,400	2,160	3,114

TABLE 3

Percent amino acid composition of acid hydrolysis of various cuticles of <u>Tenebrio</u> <u>molitor</u>. (See text for details.)

	Type of cuticle					
	Larval-		Pupal-	Adult	JH treatment	
Amino Acid	Pupal	Pupal	Adult	(1-2 day)	Outer	Inner
Aspartic Acid	5.12	4.86	7.2	5.7	3.74	5.0
Threonine	2.63	2.63	2.39	2.7	1.25	2.99
Serine	4.55	3.29	2.13	2.6	0.42	3.66
Glutamic Acid	7.00	4.60	9.8	6.7	9.76	4.73
Proline	9.00	9.1	9.53	5.2	11.4	9.22
Glycine	21.2	12.8	16.7	29.6	18.3	14.85
Alanine	11.5	26.0	14.2	10.5	19.75	24.53
Cysteine	-	-	-		2.89	-
Valine	10.1	10.8	10.1	13.2	10.8	10.78
Methionine	Trace	-	0.3	-	0.19	-
Isoleucine	4.42	5.72	4.05	4.3	4.57	3.85
Leucine	5.55	7.10	6.90	8.1	6.85	6.71
Tyrosine	3.64	2.50	4.89	1.71	3.53	2.42
Phenylalanine	1.82	1.84	2.26	1.75	2.08	1.54
Lysine	2.49	2.63	3.39	2.43	3.72	2.31
Histidine	3.45	3.54	3.28	3.4	2.48	3.52
Arginine	1.73	3.39	2.76	2.3	2.58	2.75

TABLE 4

The regulatory function of brain (Br⁺ with, Br⁻ without) in the
career of specific blood proteins, 2 weeks following perfusion
with insect Ringer, blood and a specific blood protein preparation.
The changes after treatment are noted in the same insect by comparing the composition of the blood protein before treatment. (See
text for details.)

Band No.	4.5% gel	5% gel	(no perfusion) Br^+	Br^-	Ringer Br^+	Br^-	Blood Br^+	Br^-	Protein Br^+	Br^-
1	2	2	+	0	±	±	+	±	0	-
2	5	5	+	0	±	±	±	--	0	-
3	18	20	++	0	±	-	+	--	+	--
4	25	25	--	0	-	-	++	-	-	-
5	30		--	0	0	0	--	+	+	+
		32			--	--			*	
6	38		+	0	0	-	+	+	-	-
7	48	40	0	0	0	0	0	0	0	+
% of original protein after 2 weeks			100	89	38	12	89	112	38	47

+ = increase, ± = doubtful change, 0 = no change, - = decrease,
-- = appreciable decrease, * = position of a new band

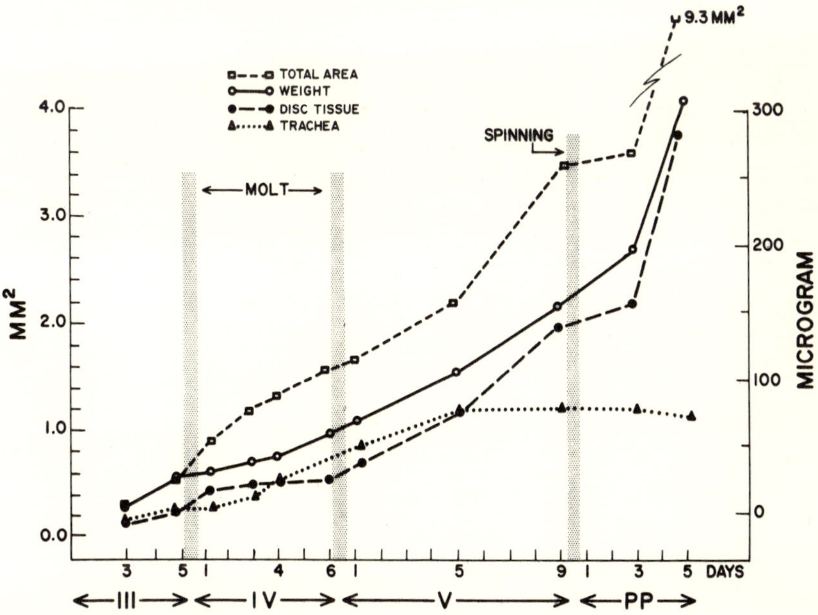

Fig. 1 Changes in the weight and size of the
wing discs of _Samia cynthia ricini_ during III, IV, V
larval instars and pharate pupal development.

183

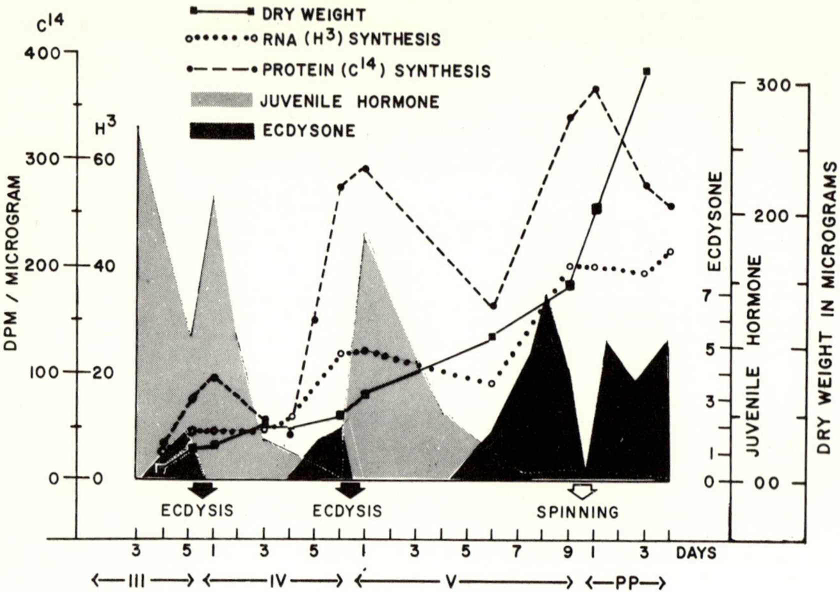

Fig. 2   RNA and protein synthetic capacities of _Ricini_ silkworm wing discs during specific stages of larval and pharate pupal development in relation to jevenile hormone, ecdysone, and the growth of the discs (dry wt.).   JH titre is expressed as _Galleria_ units/g live wt. and the ecdysone titre is expressed as _Calliphora_ units/g live wt.   RNA and protein synthesis are computed on the basis of dry weight of the wing discs.

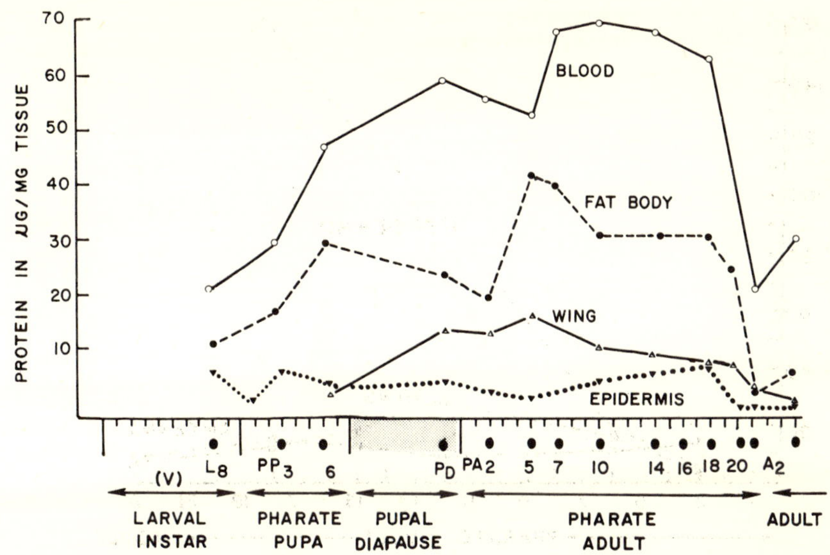

Fig. 3 Changes in the amount of total proteins (µg per mg wet weight) of blood, fat body, wings, and epidermis of <u>Hyalophora</u> <u>cecropia</u> during larval-pupal-adult development. The scale indicates the number of days between the morphologically distinct stages of IV and V instars, L-L ecdysis, L-P ecdysis, pupation, and adult emergence. Solid dots indicate specific days when experimental observations were taken, e.g., L_8 = 8-day-old V instar larva; PP_3 and PP_6 = pharate pupae (PP) 3 and 6 days respectively following the onset of spinning; PD = diapausing pupa; 2, 5, 7, 10, 14, 16, 18, 20, and 21 indicate the age of pharate adults in days from the day adult development was initiated; and A represents the 2-day-old adult.

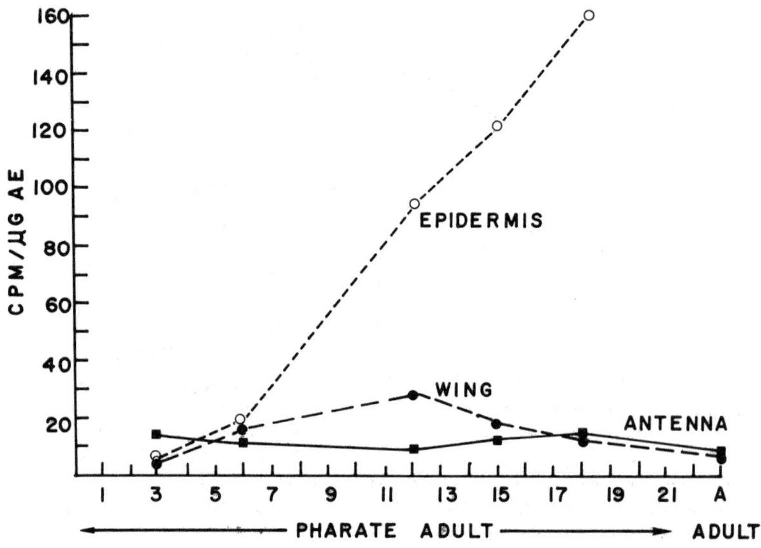

Fig. 4A Incorporation of ^{14}C-labeled amino acids into the soluble proteins in ectodermal tissues, e.g., epidermis, antennae, and wings.

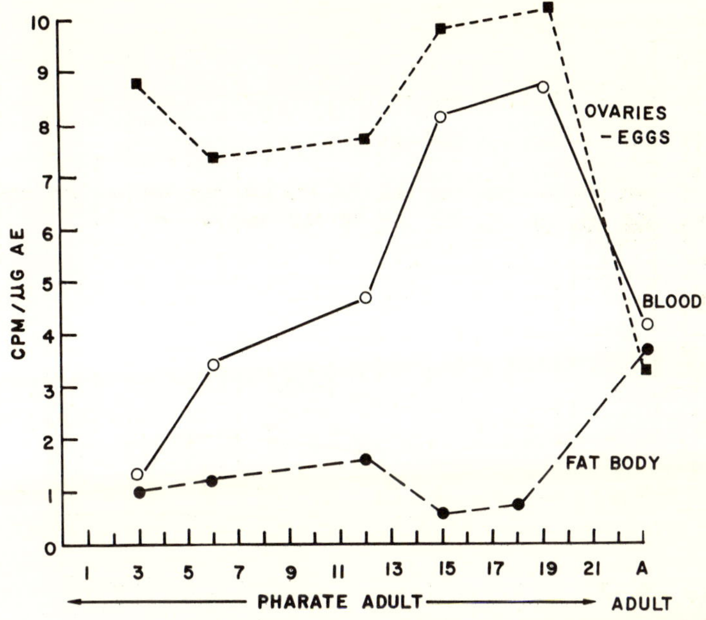

Fig. 4B Incorporation of ^{14}C-labeled amino
acids into the soluble proteins in blood, fat body,
and ovarian tissues + eggs complex. ^{14}C-labeled
amino acid precursors were injected into 18 pupae
(chilled, diapause) on day 0 of the pupal-adult
transformation. Three animals were sacrificed on
days 3, 6, 12, 15, and 18 of the pharate adult stage
and the 1-2-day adult. Tissues from all 3 animals at
each stage were used for total soluble proteins and
the data from each individual animal were similar to
others of the same stage.

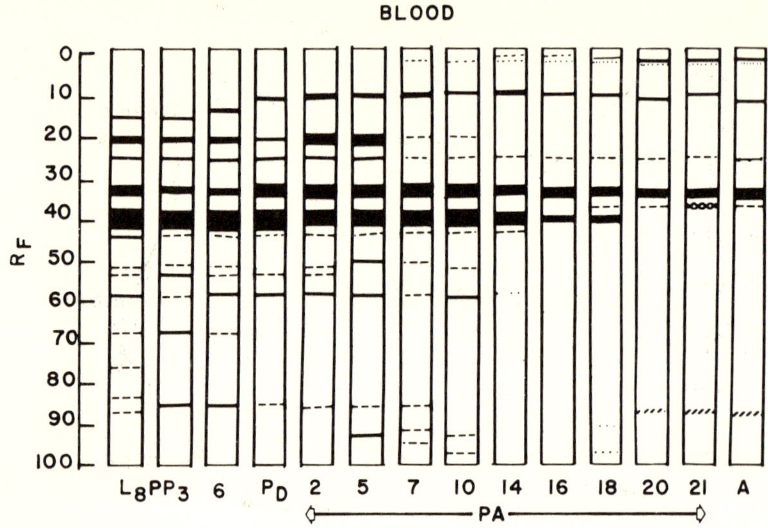

Fig. 5A   A diagrammatic representation of acrylamide gel electrophoretic separation of blood proteins of <u>H</u>. <u>cecropia</u>.   The gels represent specific stages of development of the life cycle, labeled as in Fig. 3.

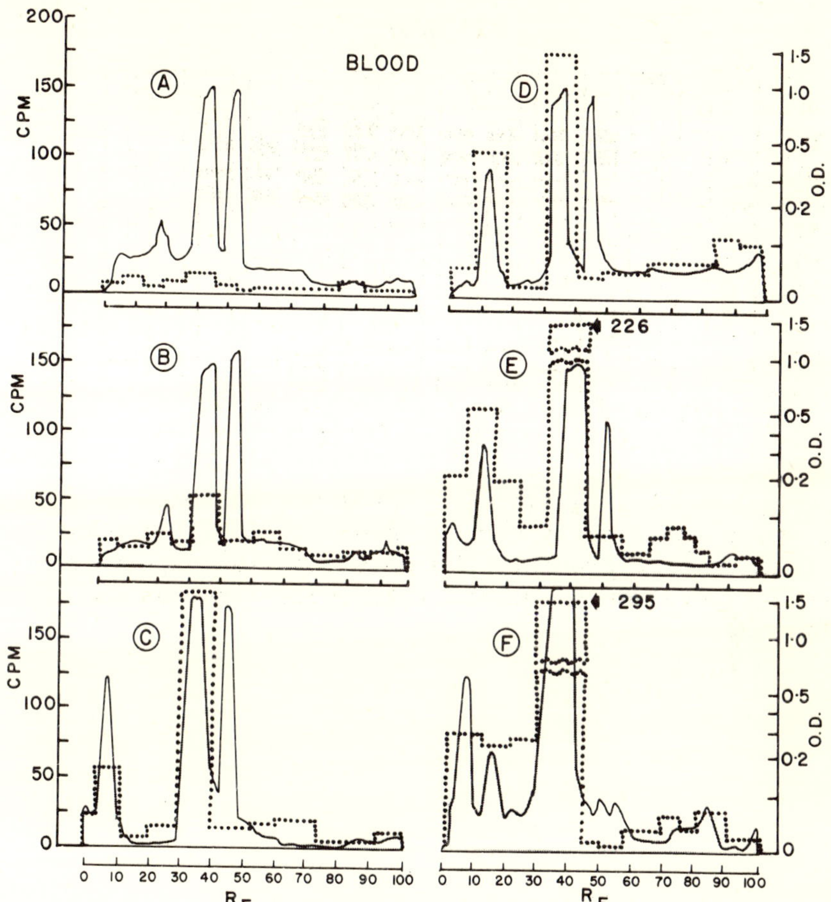

Fig. 5B Densitometric tracings (extinction, O.D.) of acrylamide gel electrophoretic separation and incorporation of labeled precursors (dotted lines) into specific blood proteins during specific stages of P-A transformation. The charts A, B, C, D, E, and F represent 3-,6-,12-,15-,and 18-day-old pharate adults and 1-day-old adult females, respectively.

FAT BODY

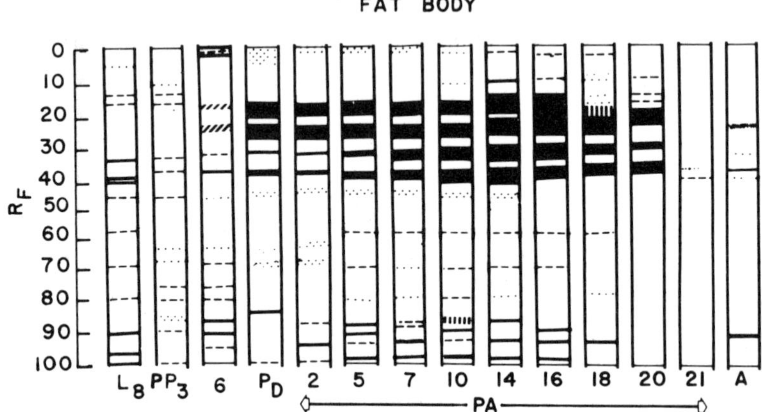

Fig. 6A   A diagrammatic representation of acrylamide gel electrophoretic separation of fat body proteins of H. cecropia.   The gels represent specific stages of development of the life cycle, labeled as in Fig. 3.

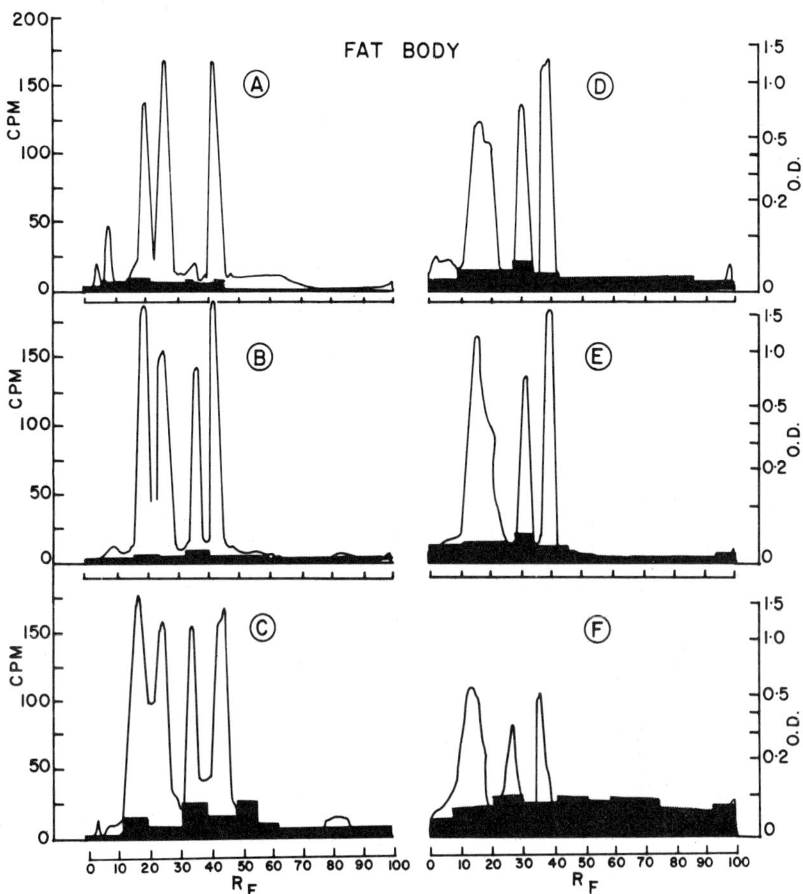

Fig. 6B Densitometric tracings (extinction, O.D.) of acrylamide gel electrophoretic separation and incorporation of labeled precursors (shaded areas) into specific fat body proteins during specific stages of P-A transformation. Other details are as in Fig. 5B.

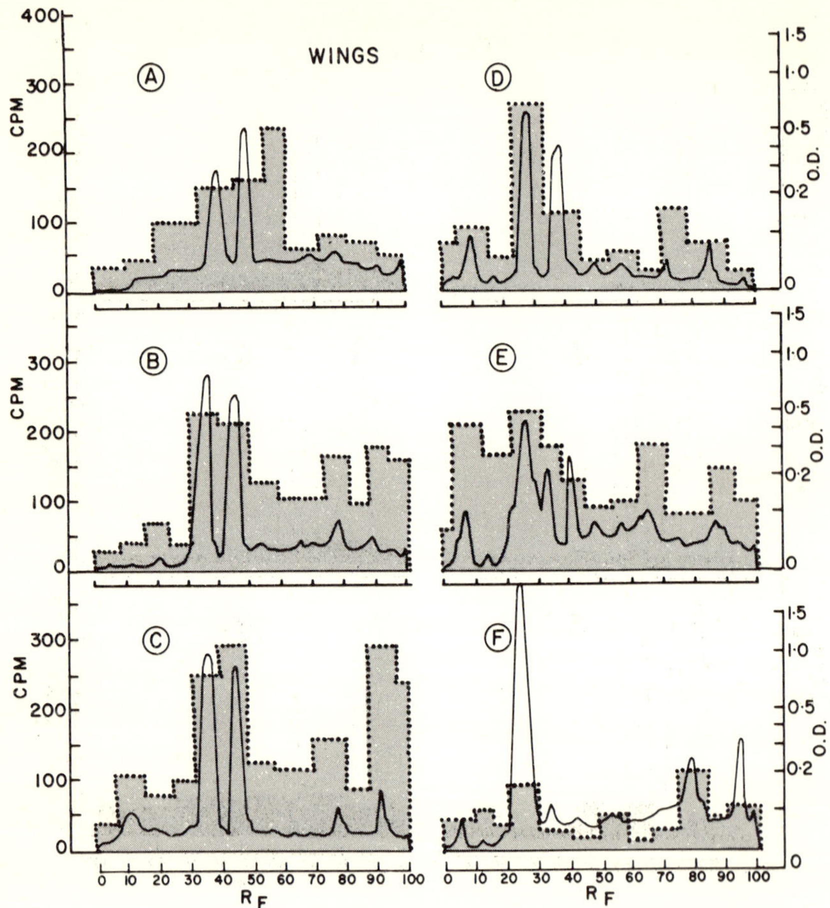

Fig. 7 Densitometric tracings (extinction, O.D.) of acrylamide gel electrophoretic separation and incorporation of labeled precursors (dotted lines) into specific wing proteins during specific stages of P-A transformation. Other details are as in Fig. 5B.

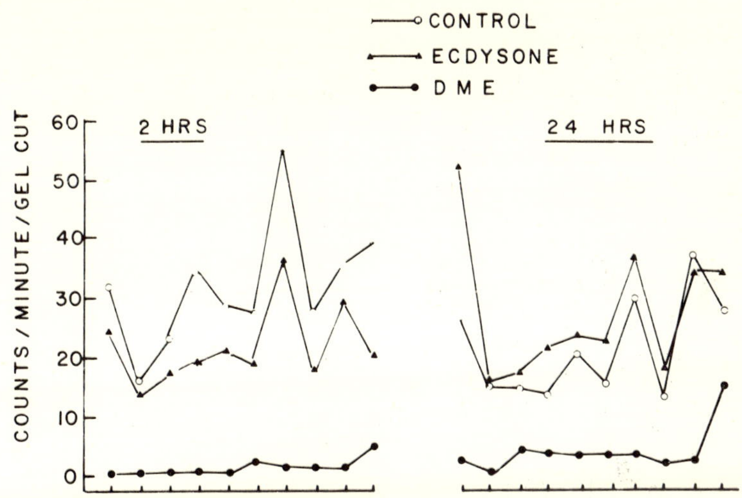

Fig. 8 Effect of DME and ecdysone on protein synthesis in <u>Tenebrio</u> <u>molitor</u> <u>in</u> <u>vivo</u> in 0-2 and 24 hrs. old pupae. The horizontal axis represents top (right end) and bottom (left) of the acrylamide gel in relation to the isotope incorporation on vertical axis.

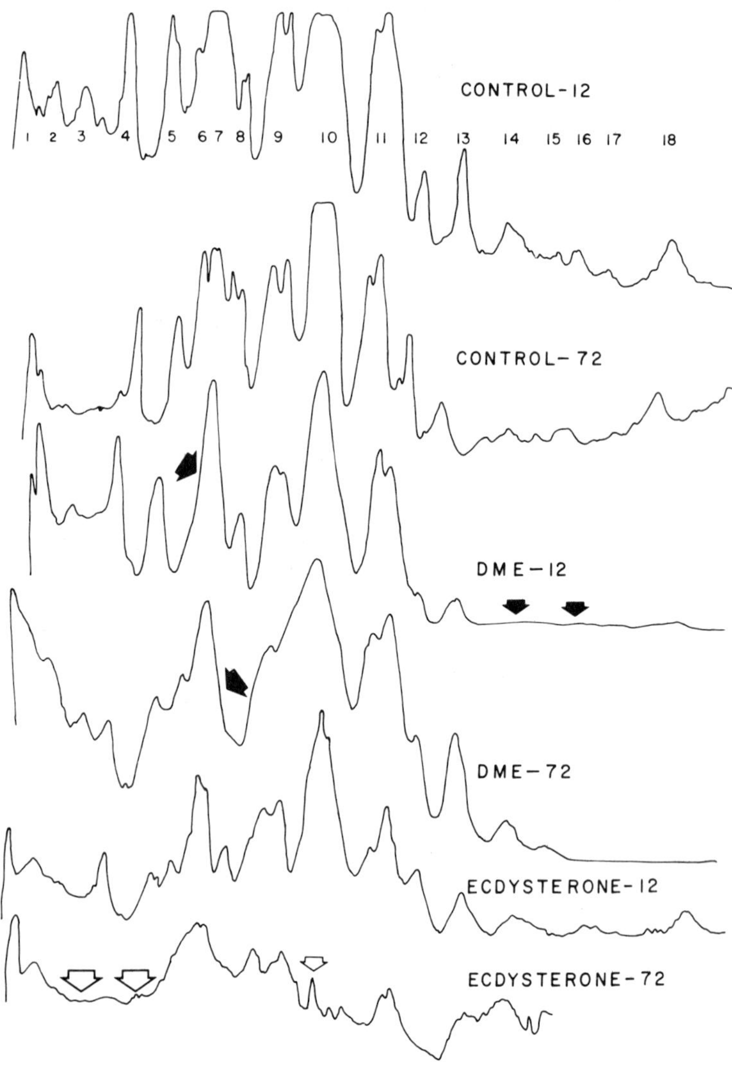

Fig. 9   Densitometric tracings of acrylamide gels of ribosomal proteins of Control, DME (1 µl/pupa) and ecdysterone (10 µg/pupa) treated _T. molitor_ of 12 and 72 hrs. old stages during P-A development. The arrows point to bands exhibiting differences.

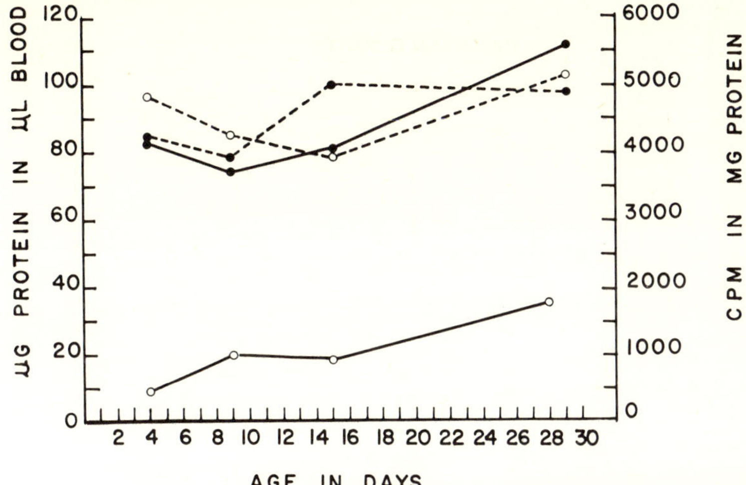

Fig. 10A The curves represent amount of total blood proteins and ^{14}C amino acid incorporation into proteins during 29-day period following ligation. The amount of protein is depicted in controls (o---o) and ecdysone (●---●), and synthesis in controls (o——o) and ecdysone (●——●).

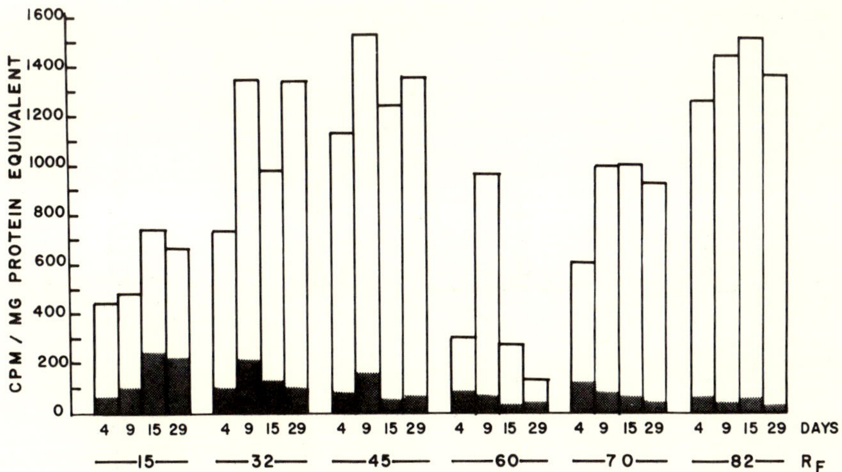

Fig. 10B A diagrammatic presentation of ^{14}C amino acid incorporation in specific blood proteins in ligated _Galleria_ _mellonella_ larvae. Each group of bars on the diagram represents a protein band separated by gel electrophoresis with designated R_F, the numbers (4, 9, 15, and 29) below each bar indicate days after the ligature. The shaded bars represent controls and the unshaded ecdysone treatment.

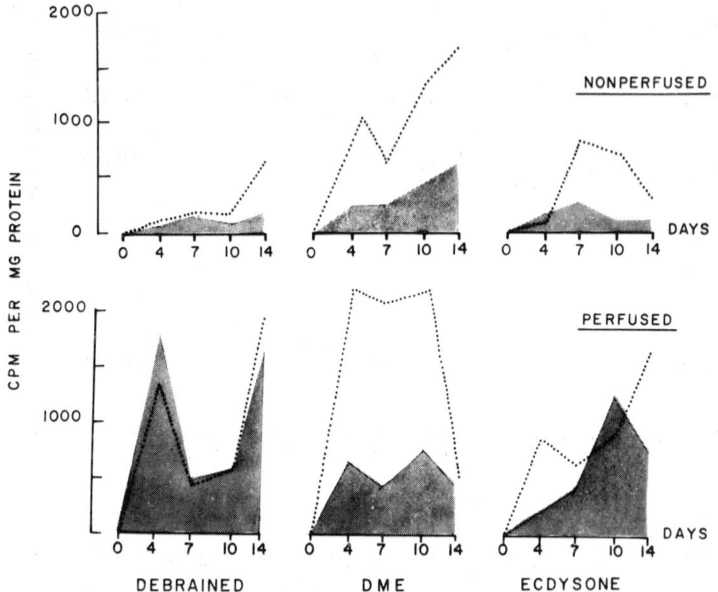

Fig. 11   The protein synthesis in non-perfused and perfused $\underline{H}$. $\underline{cecropia}$ during 0-14 day P-A development in debrained, and DME and ecdysone treated animals.   The shaded areas show changes in total proteins and the dotted lines in a specific band containing vitellogenin.

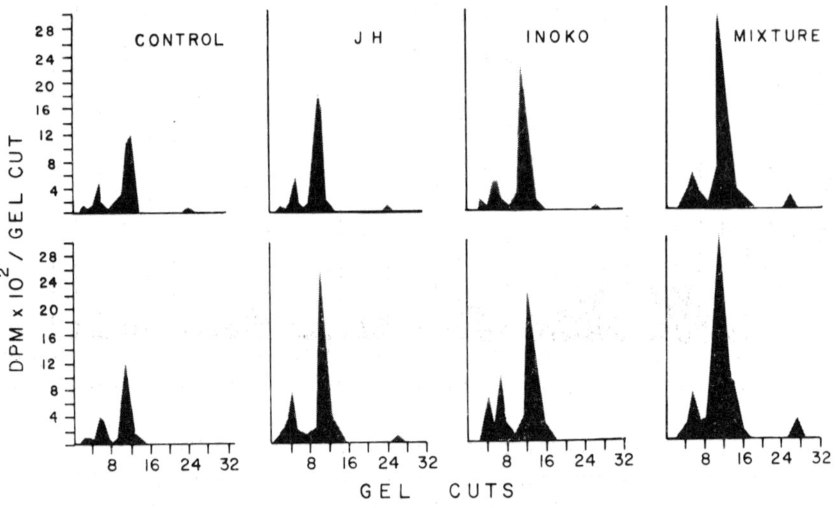

Fig. 12   The oocyte protein synthesis in 2 groups of 2-day-old $\underline{H}$. $\underline{cecropia}$ adult female, that received JH, inokosterone and their mixture at the beginning of P-A development.   The horizontal axis represents proteins separated on acrylamide gels and the vertical axis radioactivity.

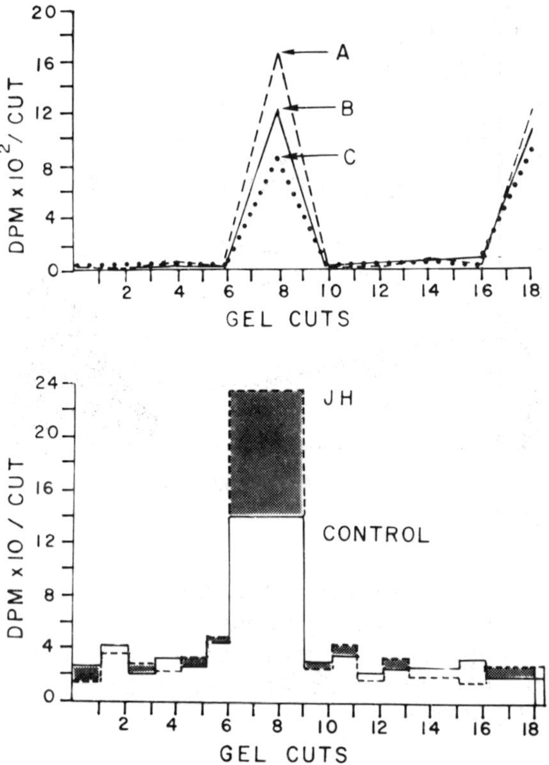

Fig. 13 *In vitro* uptake of purified vitellogenin by H. *cecropia* oocytes exposed to JH. Upper figure shows the radioactivity in the incubation medium before (A), and after (B, control and C, JH) incubation. The lower figure shows radioactivity in proteins in oocytes incubated for 4 hrs. in above medium. The horizontal axis shows position of specific protein in the gel and vertical the radioactivity.

197

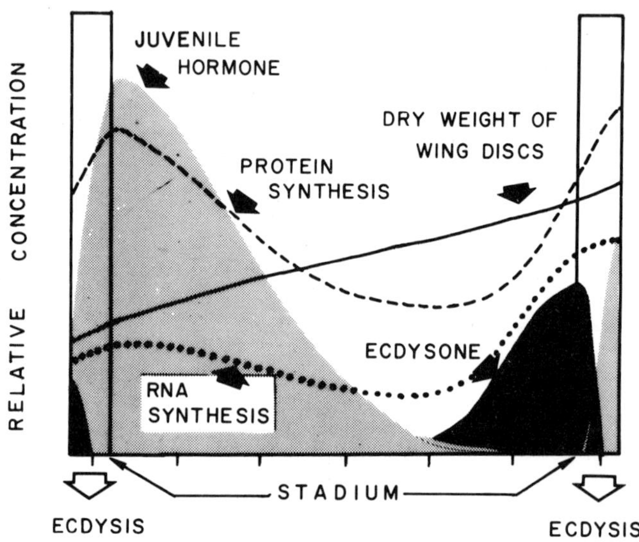

Fig. 14 A diagrammatic summary of the relationship of JH and ecdysone concentrations, their influence on the capacity of wing discs to synthesize RNA and protein, and their relationship to the rate of growth of the discs.

AMINO ACID PATTERN AND RATE OF
PROTEIN SYNTHESIS IN AGING <u>DROSOPHILA</u>*

P. S. Chen

Institute of Zoology
University of Zurich
Zurich, Switzerland

Recent research into the process of aging has
been extended to the molecular level, including alterations in the synthesis of nucleic acids, enzymes,
as well as structural proteins. There is a large
volume of data from work done mainly on mammals.
<u>Drosophila</u> has also been used by various investiga-
tors to study the problems of aging, apparently because of its short life-span and its ease of being
raised under controlled conditions. A further advantage is that the genetics of this insect are thoroughly known. The hitherto results have been
adequately discussed in a recent review by Lints
(1971).

In connection with our studies on the biochemical properties of lethal mutants in <u>Drosophila</u> (Chen,
1962, 1966), experiments dealing with changes of the
free amino acid composition and the rate of protein
synthesis during adult life have been carried out by
us. In view of the metabolic importance of these
components, the findings of this type of study can
provide us with valuable information on the basic
events involved in both development and aging.

As pointed out in a previous review (Chen,1971),
the adult life of insects can be divided into an

---

*This study was supported by grants from the
Swiss National Science Foundation and the Georges
und Antoine Claraz-Schenkung.

early period which is characterized by postemergence
differentiation and sexual maturity, and a later
period during which the aging process proceeds gradually and results ultimately in the death of the fly.
It is evident that no sharp limit can be drawn between these two periods, and the duration of each may
vary from species to species. For example, the
period of postemergence maturation lasts about only
one day in the house fly, but 10 days in the worker
honey bee (cf. Clark and Rockstein, 1964), or even
15 days in the house cricket (Nowosielski and Patton,
1965). As will be shown below, our data are in
accordance with such a functional fluctuation during
adult life.

I. Changes in free amino acid pattern during early
    adult life.

In these experiments, adult females and males of
the wild type (Sevelen) of _Drosophila_ _melanogaster_
were raised on standard medium (cornmeal-agar-sugar-
yeast) at 25°C. At the time of emergence, the culture bottles were inspected at 2-3 hour intervals,
and newly hatched flies were transferred to fresh
bottles and maintained at the same temperature. At
the desired age, the flies were etherized, separated
according to sex and preserved in the deep freezer
(-27°C) for further chemical analysis. Virgin females were also collected within 2-3 hours after
emergence and kept under the same conditions, however,
without the males.
For extraction of the amino acids, 15 female or
male flies were homogenized in 80% methanol and centrifuged. The precipitate was twice washed with
methanol and the pooled supernatant was transferred
to a Whatman No. 1 filter paper (24x46 cm). The
sheets were run in 70% n-propanol in the ascending
direction and afterwards in water-saturated phenol
in the descending direction. For quantitation of
each amino acid, the chromatogram was first sprayed
with 0.5% ninhydrin in acetone. After developing the

color for 30 minutes at 60°C, the spots were cut out
and dipped individually in a dilute $Cu(NO_3)_2$ solution, and the color intensity of the ninhydrin-
copper-complex was estimated at 510 nm, as previously described by Chen and Diem (1961).

Changes in the contents of several representative amino acids during the first week of adult life
of males as well as virgin and fertilized females
are presented in Figs. 1-4 (Chen and Geiger, unpublished). For the majority of amino acids (α ALA,
LEU/ISO, GLY, γ ABA, GLU-NH_2, THR, GLU, HIS) the
concentration appears consistently higher in females
than in males (Figs. 1-3). Determination of the wet
body weights indicated that at emergence the average
value of females amounts to 1 mg/fly, and that of
males 0.9 mg/fly. However, from the third day, body
weight increases to 1.4-1.5 mg/fly in females, whereas the corresponding value of males remains at the
level of 0.8 mg/fly. Lints and Lints (1968) also
reported that the weights of hybrid adult females of
D. melanogaster increase with age. This increase is
apparently related to growth of the ovaries in female
flies. Again in agreement with our data, Burcombe
and Hollingsworth (1970) showed that hybrid males of
D. melanogaster and D. subobscura do not change in
weight with the advance of age. Thus, the concentration difference in these amino acids is at least
partly due to differences in their body sizes.

Of further interest is the rapid increase in
different amino acid concentrations during the first
2 days after adult emergence and the subsequent decline until day 3. Furthermore, the data show clearly that the increase is more significant and reaches
a higher level in mated than in virgin females.
Several lines of evidence indicate that this difference is related to the process of egg production.
First, as is known, oviposition in mated Drosophila
females begins on the second day of adult life and
remains high compared to that of the virgin females.
Previous work suggests that the stimulation of
fecundity must be due to the introduction of the

paragonial substance into the female by the male fly
during copulation (Garcia-Bellido, 1964; Leahy, 1967;
Chen and Bühler, 1970). Secondly, the amino acids
in question, notably GLU, α ALA, GLY, and HIS, have
been found to be highly concentrated in the egg of
this insect (Chen et al., 1967). Thirdly, in contrast to male flies, there is a rapid increase in
the protease activity in the midgut of females at
the onset of adult life, apparently due to a large
demand for amino acids for the synthesis of yolk
proteins (Waldner-Stiefelmeier, 1967). Finally, as
demonstrated recently by Mohan (1971), mating in
Drosophila leads to a stimulation of ribosomal RNA
synthesis in females.

The above explanation, however, does not apply
to the male flies which exhibit likewise a distinct
increment in their amino acid contents on the first
day after emergence, though to a lesser extent in
comparison to the females. These drop again on the
second to third day and remain fairly constant until
day 7 (see Figs. 1-4). These early variations are
more likely associated with the postemergence maturation which continues at the onset of adult life.
Similarly, Nowosielski and Patton (1967), reported
that in Acheta domestica the total amino acid decreases by about 20% during the first 10 days after
emergence. Lang and associates (Lang and Stephan,
1967; Lang, 1967) observed that the total RNA/DNA
ratio in the mosquito Aedes aegypti rises in the newly hatched adults. Working with locust, Schistocerca
gregaria, Heslop (1967) found that the incorporation
of valine into wing protein is elevated from 0.17 to
1.27 µCi/g wet weight within the first 53 hours after
the last moult, decreases to 0.28 µCi/g during the
subsequent 3 weeks, and, thereafter, remains at this
low level. In D. subobscura, Maynard Smith et al.
(1970) showed that the total protein of male adults
drops rapidly during the first 4 days after emergence, and this decline is connected with the breakdown of the pupal fat body. The initial decrease of
uric acid in male adult Drosophila has also been

interpreted as due to the excretion of this substance which had accumulated during the pupal stage (Burcombe and Hollingsworth, 1970). All these facts suggest that adult differentiation is not complete at the time of emergence, and its continuity during early adult phase is reflected in changes of a variety of metabolic processes.

The patterns of VAL/MET and β ALA deserve special comments. As illustrated in Fig. 4, even if the data are expressed per unit body weight, the differences in the concentration of these amino acids between the two sexes are still valid. For technical reasons VAL and MET were not separated on our chromatograms. However, previous work by using either the micro-biological assay (Kaplan et al., 1958) or column chromatography (Chen and Bühler, 1970), all showed that female adults of Drosophila contain twice as much free MET as the males. The present data are in perfect agreement with these findings. The higher content of MET in the female fly can again be interpreted by its close relationship to egg maturation. By using autogenous and unautogenous mosquitoes of the Culex pipiens complex Chen and co-workers (Chen, 1958; Geiger, 1961; Briegel, 1969) demonstrated that the accumulation of methionine sulphoxide, an oxidative product of methionine, is correlated with ovarian development in the female adults. Moreover, in both Aedes aegypti (Dimond et al., 1956) and Anthonomus grandis (Vanderzant, 1963) it was found that MET stimulates egg-laying in females reared on synthetic medium.

In contrast to all other amino acids heretofore described, β ALA appears to be consistently more concentrated in male than in female flies, except at emergence (see Fig. 4b). It shows a 6-fold increase within the first 3 days and then drops slightly in the subsequent 4 days. In our earlier study on Culex pipiens we also found that β ALA is much more concentrated in male than in female mosquitoes (Chen, 1958). The morphogenetic and metabolic significance of such a sexual dichotomy is not clear. Recent

results revealed the incorporation of this amino
acid into cuticular protein in <u>Calliphora</u> <u>erythro-</u>
<u>cephala</u> (Karlson <u>et</u> <u>al.</u>, 1969), <u>Sarcophaga</u> <u>bullata</u>
(Bodnaryk and Levenbook, 1969), <u>Lucilia</u> <u>cuprina</u>
(Birt and Christian, 1969; Gilby and McKeller, 1970;
Hackman and Goldberg, 1971) and <u>Drosophila</u> <u>melano-</u>
<u>gaster</u> (Faller, 1970). In the last insect, its in-
corporation into the pupal sheath of the color mutant
ebony is apparently reduced compared to that of the
wild type (Jacobs, 1966; Jacobs and Brubaker, 1963).
The presence of both a β ALA decarboxylase (Jacobs
and Brubaker, 1963) and a β ALA oxidase (Duffy, 1968)
has also been demonstrated. It is further involved
in the synthesis of the dipeptide carnosine in the
blow fly larvae (Levenbook, 1966; Bodnaryk and Levenbook, 1968) and sarcophagine in the blow fly larvae
(Levenbook <u>et</u> <u>al.</u>, 1969; Bodnaryk and Levenbook,
1969). Thus, β-alanine appears to be metabolically
more active than the previous literature would seem
to indicate.

II. Effect of aging on free amino acids and related
     compounds in male adult flies.

In order to avoid the complication involved in
egg production, only male flies were used in our
study of the effect of aging on free amino acids and
protein synthesis. The flies were raised under the
same conditions as those mentioned previously, and
the amino acids were analyzed by both paper and
column chromatography. In the latter case, pooled
methanolic extracts from flies of the desired age
were taken to dryness in a flash evaporator, and the
dry residue was dissolved in redistilled water to a
final concentration of 0.1 g/ml. The resulting
solution was then shaken with an equal volume of
chloroform. Subsequent to separation of the two
layers by standing, the aqueous solution was care-
fully pipetted out and, if necessary, again centrifuged. Fractionation of all ninhydrin-positive components was done by an automatic amino acid analyzer

(Beckman 120 B) following the procedures given by
Spackman et al. (1958). The identification of individual peaks has been described by Chen and Hanimann
(1965)

In the present experiment we used males aged 0,
3, 20, and 30 days. Since the viability of the flies
decreases in the course of aging, these had to be
transferred to fresh culture bottles every 4-5 days.
Despite such a precaution our experience indicates
that the female flies are much more sensitive than
the males, so that at the later periods male flies
usually became predominant in each bottle.

The quantitative data from paper chromatography
are summarized in Table 1 (Chen and Stäuble, unpublished). Owing to the limit of resolution only those
components which were distinctly separated on the
paper chromatogram were estimated. In agreement with
the results illustrated in Figs. 1-4, most of these
amino acids show the lowest concentration by day 3.
This is also clearly shown by the total concentration. We shall discuss in more detail the variations
of individual components later.

The analyzer profiles of adult males aged 0 and
30 days are depicted in Figs. 5 and 6 respectively.
In addition to the acidic components eluted before
ASP (see below), the high concentration of β ALA and
the low concentration of LYS in the 30-day-old flies
are quite impressive.

Table 2 includes the quantitative changes of all
ninhydrin-positive compounds as determined by the
amino acid analyzer. As can be seen, there is apparently no significant variation of the total concentration during the 30 days of adult life-span. This
is in agreement with the results of Burcombe and
Hollingsworth (1970) from their study on the aging
male hybrids of D. melanogaster and D. subobscura.
The differences between the data presented in Table 1
and Table 2 are because in the paper chromatographic
procedure only a limited number of amino acids were
selected for quantitation.

As to the individual components, both Tables 1

and 2 show that β ALA has a nearly 3-fold increment
during adult life, so that it becomes one of the
most concentrated amino acids in the 30-day-old
flies. This is true whether the data are expressed
as μg/insect or as unit wet weight or as % total.
GLY is maintained at an essentially constant level,
whereas LYS declines regularly in the course of
aging. By contrast, α ALA increases distinctly from
day 0 to day 20-30. We shall refer to these amino
acids again in Section III when dealing with the
rate of protein synthesis (p. 9).

Furthermore, studies with the amino acid analyzer disclosed significant fluctuation of various
acidic components (Fig. 7). The paragonial substance
(PS) is not detectable at emergence, rises rapidly
to a prominent peak within 3 days, and is kept at an
extremely high level until 20-30 days. That the
early increase occurs at the most active reproductive
period is understandable. Its high level at later
periods is possibly related to the progressive elim-
ination of females in the culture bottles with advancing age. Unpublished work in our laboratory
has shown that in the absence of mating, i.e., when
the male flies are separated from females, there is
an accumulation of this substance in the paragonial
glands.

The phosphate ester α-glycerophosphoethanolamine
(GPEA) exhibits also a rapid increase at day 3, but
falls off again at day 20, and becomes hardly detectable until day 30.

Throughout the 30-day experimental period, both
phosphoethanolamine (PEA) and taurine (TAU) occur
regularly as two prominent peaks, but at all periods
so far examined the concentration of taurine remains
proportionally higher.

These results demonstrate clearly that there are
distinct age-dependent changes of individual ninhydrin-positive components, even though their total
concentration remains largely constant (see Burcombe
and Hollingsworth, 1970 and Table 2 in this study).

In the mosquito _Aedes aegypti_ Stidham and Liles

(1969) found that adult females maintained on either sucrose or sucrose and blood have a highest total amino acid concentration at 30 days of age. α ALA, which comprises 34-52% of the total concentration, shows a similar pattern. From their statistical analysis of the data it has been concluded that diet had a significant effect on individual amino acids only when age was also considered. Working with the same insect Thayer and Terzian (1962, 1970) reported that many amino acids in adult female mosquitoes fed with sucrose alone declined during the first 7 days after emergence, probably reflecting postemergence development as has been suggested for _Drosophila_. Their data show that during the period of senescence, most amino acids are maintained at a rather constant level, whereas some others rise temporarily. Without detailed information on the metabolic fate and the overall utilization of individual components an interpretation of such transitional changes is by no means easy.

There is no doubt that in order to achieve maximal longevity an adequate diet has to be supplied. In both _Calliphora_ (Tribe, 1966) and _Drosophila_ (Hollingsworth and Burcombe, 1970) it has been shown that sugar, a respiratory substrate, is necessary for adult maintenance. However, the essential amino acids must be derived from yeast or other protein sources.

III. Effect of aging on the rate of protein synthesis in male adult flies.

Clarke and Maynard Smith (1966) claimed that in _Drosophila_ _subobscura_ the incorporation rate of tritiated leucine into protein is twice higher in male adult flies at 60 days than those at 20 days of age. Similar results were found for the incorporation of labeled uracil into RNA. They suggested that the elevated rate is related to a greater loss of body protein in the aged flies. The results of Clarke and Maynard Smith are, however, in contradic-

tion to that reported by Heslop (1967) who observed
no age-dependent increase in the incorporation of
labeled valine into wing protein of the locust
Schistocerca gregaria. In fact, the more recent
study of Levenbook and Krishna (1971) on the blow
fly Phormia regina and our own study on Drosophila
melanogaster (Baumann and Chen, 1968; Baumann, 1969)
pointed out that the opposite is true.

In the previous sections we have seen that the
individual free amino acid pool in young and old
flies may differ considerably. For this reason we
decided to use kinetic analyses to estimate the rate
of protein synthesis. Flies aged 3, 20, and 50 days
after emergence were individually injected with
uniformly ^{14}C-labeled lysine, glycine or α-alanine,
and samples were taken at 0, 30, 60, 90, and 120 min.
intervals. The amount of amino acid solution injected was about 0.7-1.0 mCi/µl (specific activity 180
mCi/mM for lysine, 108 mCi/mM for glycine and 85
mCi/mM for alanine; Radiochemical Centre, Amersham).
The reason for employing these amino acids as markers
is that lysine is metabolically inactive and essential for Drosophila, whereas α-alanine is metabolically active and non-essential. According to Hinton
et al. (1951) glycine seems to be important for
growth of Drosophila larvae. In addition, these
three amino acids can be distinctly separated by one-
dimensional high-voltage paper electrophoresis at
pH 1.5 (8% formic acid) and 2500 V (see Chen et al.,
1968).

Owing to the small size, four flies had to be
used for each determination. The flies taken at the
desired interval after injection were homogenized in
a conical microhomogenizer in 300 µl 0.3 N perchloric
acid (PCA) and heated for 2 min. at 100°C. Subsequent to centrifugation the precipitate was washed
three times with the same volume of PCA. The pooled
supernatant was then neutralized with 1.5 N KOH.
After drying at 80°C the amino acids were washed out
from $KClO_4$ with 80% ethanol. Half of the sample was
used for determining the pool size and the other half

for estimation of the radioactivity of the amino acid in question.

The protein precipitate from PCA was at first washed twice with ether-ethanol (1:3) and then dissolved in concentrated formic acid. About 10% of this sample served for the determination of total radioactivity, 60% for total protein, and 30% for radioactivity of the amino acid incorporated into protein. In general, following electrophoretic separation, the amino acid concentration was determined by the cadmium-ninhydrin-reagent (Atfield and Morris, 1961) and its radioactivity by a gas-flow counter (Nuclear Chicago). The biuret reagent was used for estimating the total protein.

On the basis of parameters obtained by the above procedures, the turnover rate of the injected amino acid (k_a) and its rate of incorporation into total protein (K_p) were calculated by the equations of J.Z. Hearon as published by Dinamarca and Levenbook (1966). In addition, we determined the oxidation rate of the injected amino acid (K_p) according to equations derived in this laboratory (see Baumann, 1969). It should be pointed out that all these equations were deduced by assuming that there is only one homogeneous pool, i.e., a complete mixing of the amino acids in the pool, and that there is only a negligible recycling or return of label to the pool. These assumptions apparently do not correspond to the real situation since evidence from our recent study on the biosynthesis of amino acids from ^{14}C-glucose in _Drosophila_ indicates that the free pool may not be identical with the pool available to protein synthesis (Widmer, 1972). Furthermore, work on rat liver (Gan and Jeffay, 1967) and Hela cells (Righetti _et al_., 1971) have shown that amino acids from protein and peptide breakdown flow back to the pool and are reutilized for protein synthesis. Nevertheless, in view of the complex situation during aging, it seems to us that a meaningful study of protein synthesis is possible only when the dynamic state of the free amino acids is considered.

As illustrated in Fig. 8, a plot of K_a as a function of time yields linear curves, the slopes of which correspond to the turnover rates of flies aged 3, 20, and 50 days. The same is true for the incorporation rate K_p. The results are in line with the proposed first-order kinetics in the derivation of the equations.

The data of total free pools and the computed turnover rates for the three injected amino acids are included in Table 3, Both the free lysine and glycine pools decrease from 3 to 50 days, whereas the free alanine pool increases. Likewise, the turnover rates of lysine and glycine are lower in 50-day-old as compared with 3-day-old flies, but that of alanine appears higher. However, when the latter values are expressed as % of total pool, there is a reduction in the turnover of all three amino acids with the progress of aging.

As summarized in Table 4, the incorporation rates of lysine, glycine, and alanine decrease continuously throughout the 50-day experimental period. On a per insect basis the values for the flies aged 50 days amount to only about 42-63% of that of the flies at 3 days of age. When the data are calculated on a per mg protein basis, the differences are even more impressive (37-42%).

IV. Effect of aging on the oxidation rate of amino acids.

A specially designed aluminum chamber adapted to the D34 counter (Nuclear Chicago) was used for determining the production of $^{14}CO_2$ of male flies injected with labeled lysine, glycine, and alanine (for details, see Baumann, 1969). In principle, this method is similar to that described by Levenbook and Dinamarca (1965) who employed an ionization chamber for measuring the respiratory $^{14}CO_2$ in Phormia. A total of 4 flies had to be used for each experiment.

The patterns of $^{14}CO_2$ production within 2 hours after injection for flies aged 3 and 50 days are presented in Fig. 9. It is clear that alanine is most rapidly oxidized, and lysine is least rapidly respired. Glycine behaves in an intermediate manner. For all three amino acids the highest oxidation rates are within about 30 min. after injection. It should be pointed out that the progressive curves in Fig. 9 show only the cumulative production of $^{14}CO_2$ in the respiratory chamber, and that neither the pool size nor the turnover of the injected amino acid have been taken into consideration.

The oxidation rate constants computed from the expired $^{14}CO_2$, the total pool and the turnover rates are given in Table 5. The values at 50 days are about 85-95% of those at 3 days. The difference between young and old flies is smaller than that found for the turnover and incorporation rates. It seems that the respiratory metabolism per se is no limiting factor for the reduced rate of protein synthesis in the aged flies. Furthermore, the data expressed at % pool indicate that alanine is very active, glycine fairly active, while lysine is inactive.

In general, our data on the rates of protein synthesis are in reasonable agreement with those reported for <u>Phormia regina</u> by Levenbook and Krishna (1971). They showed that the K_p in flies aged 37-40 days amounts to only 26% of that in flies aged 5-6 days for U-^{14}C-alanine, and 52% for U-^{14}C-lysine. As already noted, Heslop (1967) detected also no increase in the synthesis of wing protein with aging in the locust <u>Schistocerco gregaria</u>. Thus, the results of this type of study on three different insects are all at variance with the previous findings of Clarke and Maynard Smith (1966) for <u>Drosophila subobscura</u>. There is no reason to assume that different species differ principally in such a fundamental metabolic process like protein synthesis. Likewise, it is very unlikely that leucine behaves differently from lysine, glycine, or alanine. How-

ever, it should be pointed out that, at least as far
as our data show, the depression in protein synthesis
is slower between 20 and 50 days than that during the
first 20 days. In Clarke and Maynard Smith's experiment this early period was not included. In other
words, when only the data between 20 and 50-60 days
are compared, the difference is smaller. Another
point is that in their study on D. subobscura the
flies were fed with ^{3}H-leucine, while in ours and
that by Levenbook and Krishna (1971) the amino acids
were injected into the haemocoel. Although it has
been shown that the incorporation of the orally administered leucine into protein was not delayed by
the absorption process (Dingley and Maynard Smith,
1968), it is not known to what extent the gut epith-
elium is influenced by age. A third and more critical point is that Clarke and Maynard Smith measured
only the total pool and the total radioactivity in
protein, whereas in the experiments by us and by
Levenbook and Krishna a kinetic analysis including
the turnover rate was performed. Granting that all
other parameters remain constant, a reduced turnover
rate with aging would lead to a proportionally higher
level of radioactivity in the precursor pool, resulting in a seemingly higher rate of protein synthesis.
That this is indeed the case has been shown by our
own experience. In several experiments where the
labeling of protein was "directly" estimated, aged
flies may exhibit a more rapid amino acid incorporation than younger ones.

In conclusion, our study on Drosophila melanogaster demonstrated that both the turnover rates of
free amino acids and their incorporation into protein
definitely become depressed in the 50-day adult fly.
This is in accord with previous biochemical and
physiological studies in relation to aging in several
other insects. Rockstein and associates found that
in adult Musca domestica the content of AMP (Rockstein and Gutfreund, 1961) and the activity of ATPase,
NAD-dependent α-glycerophosphate dehydrogenase (Rockstein and Brandt, 1963) and acid phosphatases (Clark

and Rockstein, 1964) diminish rapidly between 6 and
12 days. In <u>Apis mellifera</u> there is rapid decline of
the activity of alkaline phosphatases (Rockstein,
1950). According to Lang and collaborators (Lang <u>et
al</u>., 1965; Lang and Stephan, 1967; Lang, 1967) the
RNA/DNA ratio in adult <u>Aedes aegypti</u> is about 30-40%
higher in the first 8 days than at later periods, suggesting a more active protein synthesis during early
adult life. Furthermore, the three enzymes glucose-
6-phosphate dehydrogenase, gluconic acid dehydrogenase, and isocitrate dehydrogenase show a higher activity in the first 10 days, but fall off by day 10 to
20. In <u>Calliphora erythrocephala</u> it has been found
by Tribe (1966, 1967) that the respiratory rate in
flies aged 80 days is only about 50% of that at the
first week after emergence, and the oxidative phos-
phorylation in the flight muscle mitochondria deter-
iorates with age. Working with <u>Drosophila melano-
gaster</u> hybrids, Lints and Lints (1969) also noted
that the rate of oxygen uptake diminishes during aging. All the above results are in agreement with the
generally accepted view that enzyme activity and protein synthesis decline with advancing age. The
genetic implications of the aging process have been
discussed by von Hahn (1966a,b).

SUMMARY

1. Studies on the free amino acids during early
adult life of Drosophila melanogaster showed that
α ALA, LEU/ISO, GLY, γ ABA, GLU-NH$_2$, THR, GLU, and
HIS increase rapidly during the first 2 days after
emergence and then decline until day 3. Several
lines of evidence indicate that these variations are
related to post-emergence differentiation and ovarian
growth.
2. In general, the amino acids increase more
rapidly and reach a higher concentration in mated
than in virgin female flies. This difference is
probably due to the stimulation of egg production by
the paragonial substance introduced into the female

during copulation.

    3. In accordance with previous results adult females contain twice as much free MET as the males, whereas the reverse is true for β ALA.

    4. Analyses of the free amino acids in male flies during a 30-day period of adult life revealed that the total content of ninhydrin-positive components remains essentially constant. There are, however, significant changes of individual amino acids and related compounds. Thus, compared to younger flies, adult males aged 30 days contain more TAU, α ALA, β ALA and less GLY, VAL, and LYS.

    5. A kinetic investigation of the effect of aging on the rate of protein synthesis demonstrated that there is a definite decrease in the incorporation of lysine, glycine and α-alanine into protein with age. Compared to adult males at day 3, the values for flies aged 50 days are only 37% for LYS, 42% for GLY and 38% for ALA. The oxidation rates of these amino acids suggest that respiratory metabolism is probably not a limiting factor for the reduced rate of protein synthesis.

REFERENCES

Atfield, G.N. and Morris, C.J.R. (1961). _Biochem. J._ **81**, 606.

Baumann, P.A. (1969). _Z. vergl. Physiol._ **64**, 212.

Baumann, P.A. and Chen, P.S. (1968). _Rev. suisse Zool._ **75**, 1051.

Birt, L.N. and Christian, B. (1969). _J. Insect Physiol._ **15**, 711.

Bodnaryk, R.P. and Levenbook, L. (1968). _Biochem. J._ **110**, 771.

Bodnaryk, R.P. and Levenbook, L. (1969). _Comp. Biochem. Physiol._ **30**, 909.

Briegel, H. (1969). _J. Insect Physiol._ **15**, 1137.

Burcombe, J.V. and Hollingsworth, M.J. (1970). _Exp. Geront._ **5**, 247.

Chen, P.S. (1958). _J. Insect Physiol._ **2**, 128.

Chen, P.S. (1962). _In_ "Amino Acid Pools" (J. T.

Holden, ed.), p. 115, Elsevier, Amsterdam.
Chen, P.S. (1966). Adv. Insect Physiol. 3, 53.
Chen, P.S. (1971). "Biochemical Aspects of Insect Development", Karger, Basel.
Chen, P.S. and Bühler, R. (1970). J. Insect Physiol. 16, 615.
Chen, P.S. and Diem, C. (1961). J. Insect Physiol. 7, 289.
Chen, P.S. and Hanimann, F. (1965). Z. Naturforsch. 20b, 307.
Chen, P.S., Hanimann, F., and Briegel, H. (1967). Rev. suisse Zool. 74, 570.
Chen, P.S., Kubli, E., and Hanimann, F. (1968). Rev. suisse Zool. 75, 509.
Clark, A.M. and Rockstein, M. (1964). In "The Physiology of Insecta" (M. Rockstein, ed.), Vol. I, Ch. 6, Academic Press, New York and London.
Clarke, J.M. and Maynard Smith, J. (1966). Nature 209, 627.
Dimond, J.B., Lea, A.O., Hahnert, W.F., and DeLang, D.M. (1956). Canad. Ent. 88, 57.
Dinamarca, M.L. and Levenbook, L. (1966). Arch. Biochem. Biophys. 117, 110.
Dingley, F. and Maynard Smith, J. (1968). J. Insect Physiol. 14, 1185.
Duffy, J.P. (1968). J. Insect Physiol. 14, 199.
Faller, E. (1970). Diplomarbeit, Universitat Zurich.
Gan, J.C. and Jeffay, H. (1967). Biochim. Biophys. Acta 148, 448.
Garcia-Bellido, A. (1964). Z. Naturforsch. 19b, 491.
Geiger, H.R. (1961). Rev. suisse Zool. 65, 583.
Gilby, A.R. and McKeller, J.W. (1970). J. Insect Physiol. 16, 1517.
Hackman, R.H. and Goldberg, M. (1971). J. Insect Physiol. 17, 335.
Hahn, H.P. von (1966a). J. Geront. 21, 291.
Hahn, H.P. von (1966b). 7th Int. Congr. Geront., 243.
Heslop, J.P. (1967). Biochem. J. 104, 5P.
Hinton, T., Noyes, D.T. and Ellis, J. (1951). Physiol. Zool. 24, 335.
Hollingsworth, J.J. and Burcombe, J.V. (1970). J.

Insect Physiol. 16, 1017.
Jacobs, M.E. (1966). Genetics 53, 777.
Jacobs, M.E. and Brubaker, K.K. (1963). Science 139, 1282.
Kaplan, W.D., Holden, J.T. and Hochman, B. (1958). Science 127, 473.
Karlson, P., Sederi, K.E. and Marmaras, V.K. (1969). J. Insect Physiol. 15, 319.
Lang, C.A. (1967). J. Geront. 22, 53.
Lang. C.A. and Stephan, J.K. (1967). Biochem. J. 102, 331.
Lang, C.A., Lau, H.Y. and Jefferson, D.J. (1965). Biochem. J. 95, 372.
Leahy, M.G. (1967). J. Insect Physiol. 13, 1283.
Levenbook, L. (1966). Comp. Biochem. Physiol.18, 341.
Levenbook, L. and Dinamarco, M.L. (1965). Analyt. Biochem. 11, 391.
Levenbook, L. and Krishna, I. (1971). J. Insect Physiol. 17, 9.
Levenbook, L., Bodnaryk, R.P. and Spande, T.F. (1969). J. Biochem. 113, 837.
Lints, F.A. (1971). Gerontologia 17, 33.
Lints, F.A. and Lints, C.V. (1968). Exp. Geront. 3, 341.
Lints, F.A. and Lints, C.V. (1969). Exp. Geront. 4, 81.
Maynard Smith, J. Bozcuk, A.N. and Tebbutt, S. (1970). J. Insect Physiol. 16, 601.
Mohan, J. (1971). J. Insect Physiol. 17, 1061.
Nowosielski, J.W. and Patton, R.L. (1965). J. Insect Physiol. 11, 263.
Righetti, P., Little, E.P. and Wolf, G. (1971). J. Biol. Chem. 246. 5724.
Rockstein, M. (1950). J. Cell Comp. Physiol. 35, 11.
Rockstein, M. and Brandt, K.F. (1963). Science 139, 1049.
Rockstein, M. and Gutfreund, D.D. (1961). Science 133, 1476.
Spackman, D.H., Stein, W.H. and Moore, S. (1958). Analyt. Chem. 30, 1190.
Stidham, J.D. and Liles, J.N. (1969). J. Insect

Physiol. 15, 1969.
Thayer, D.W. and Terzian, L.A. (1962). J. Insect
    Physiol. 8, 133.
Thayer, D.W. and Terzian, L.A. (1970). J. Insect
    Physiol. 16, 1.
Tribe, M.A. (1966). J. Insect Physiol. 12, 1577.
Tribe, M.A. (1967). Comp. Biochem. Physiol. 23, 607.
Vanderzant, E.S. (1963). J. Insect Physiol. 9, 683.
Waldner-Stiefelmeier, R.D. (1967). Z. vergl. Physiol.
    56, 268.
Widmer, B. (1972). Insect Biochem. (in press).

TABLE 1

Free amino acid content (µg/20 flies) in
0-, 3- and 30-day-old male adults of Drosophila melanogaster

Amino Acid	0 day		3 days		30 days	
	µg	% Total	µg	% Total	µg	% Total
ASP	12.91	8.57	6.49	6.73	4.07	3.30
GLU	7.46	4.95	6.76	7.01	7.46	6.06
SER	2.38	1.58	2.06	2.14	2.06	1.67
GLY	5.50	3.65	3.88	4.03	4.46	3.62
TAU	4.75	3.15	4.35	4.51	7.15	5.81
LYS/ORN	8.80	5.84	4.03	4.18	3.20	2.60
ARG	20.14	13.37	13.89	14.41	19.51	15.55
GLU/NH$_2$	13.53	8.99	8.58	8.90	9.20	7.47
THR	3.21	2.13	0.97	1.01	1.20	0.97
Alpha-ALA	18.48	12.27	15.57	16.16	22.12	17.96
Beta-ALA	8.83	5.86	11.94	12.39	22.96	18.40
TYR	6.95	4.62	1.52	1.58	2.36	1.92
HIS	9.07	6.02	7.10	7.37	8.89	7.22
Gamma-ABA	10.88	7.22	4.05	4.20	3.74	3.04
PRO	2.31	1.53	0.98	1.02	0.88	0.71
VAL/MET	7.13	4.73	2.34	2.43	2.00	1.62
LEU/ISO	8.26	5.49	1.86	1.93	2.20	1.79
Total	150.59		96.37		123.16	

Average values from five determinations by 2-dimensional paper
chromatography according to Chen and Diem (1961).

TABLE 2

Distribution of free amino acids and related compounds
(µMoles/g wet weight and % total) in 0- to 30-day-old
male adults of _Drosophila_ _melanogaster_

Compound	0 day		3 days		20 days		30 days	
	µMoles/g	%Total	µMoles/g	%Total	µMoles/g	%Total	µMoles/g	%Total
PSER	0.48	0.89	0.87	1.34	0.61	0.97	0.75	1.17
TYRP	0.45	0.84	-	-	-	-	-	-
GPEA	0.54	1.01	3.26	5.04	1.78	2.84	1.41	2.20
PEA	2.47	4.59	1.97	3.04	2.35	3.75	4.35	6.80
TAU	3.22	5.99	5.74	8.87	7.39	11.80	8.22	12.85
MSO	0.07	0.13	0.06	0.09	0.02	0.03	0.06	0.09
ASP	0.08	0.15	0.05	0.08	0.05	0.08	0.19	0.30
THR	0.95	1.77	0.59	0.91	0.46	0.74	0.38	0.59
SER+ASPNH$_2$+ GLUNH$_2$	5.64	10.49	2.72	4.20	2.18	3.48	2.79	4.36
GLU	2.46	4.58	1.88	2.91	1.79	2.86	1.89	2.95
PRO	8.56	15.92	7.02	10.85	5.80	9.26	5.01	7.83
GLY	2.45	4.56	2.33	3.60	2.23	3.56	1.96	3.06
α ALA	9.60	17.86	14.67	22.67	13.88	22.17	11.61	18.15
α ABA	0.04	0.07	-	-	+	-	-	-
VAL	1.17	2.18	0.66	1.02	0.44	0.70	0.30	0.47
MET	0.23	0.43	0.41	0.63	0.05	0.08	0.28	0.44
ISO	0.49	0.91	0.51	0.79	0.26	0.42	0.26	0.41
LEU	0.66	1.23	0.56	0.87	0.36	0.58	0.22	0.34
TYR	0.71	1.32	0.04	0.06	0.16	0.26	0.17	0.27
PHE	0.36	0.67	0.20	0.31	0.11	0.18	0.08	0.12
β ALA	2.71	5.04	7.33	11.33	8.53	13.62	7.37	11.52
γ ABA	0.56	1.04	0.95	1.47	0.92	1.47	0.93	1.45
ORN	+	-	0.04	0.06	+	-	0.08	0.13
EA	0.80	1.49	1.37	2.12	0.98	1.57	0.64	1.00
NH$_3$	(1.80)	-	(2.64)	-	(1.88)	-	(1.64)	-
LYS	2.57	4.78	0.95	1.47	0.71	1.13	0.80	1.25
HIS	3.05	5.67	3.08	4.76	2.93	4.68	3.83	5.99
TRY	+	-	0.49	0.76	0.37	0.59	0.37	0.58
ARG	2.51	4.67	2.54	3.92	2.41	3.85	2.64	4.13
3M-HIS	-	-	0.11	0.17	0.18	0.29	0.20	0.31
PS	-	-	3.06	4.73	4.58	7.32	5.00	7.81
Peptide 2	0.93	1.73	1.26	1.95	1.08	1.73	2.19	3.42
Total (excluding NH$_3$)	53.76		64.72		62.61		63.98	

TABLE 3

Pool size and turnover rate of free lysine, glycine and α-alanine in male adults of _Drosophila melanogaster_ of different ages

Amino acid	Age (days)	Pool size (μMoles/fly)	Turnover rate K_a (μMoles/hr/fly	% Pool turnover/hr
Lysine-U-^{14}C	3	6.64×10^{-4}	3.29×10^{-4}	49.6
	20	6.06	2.56	42.2
	50	4.71	1.85	39.4
Glycine-U-^{14}C	3	2.26×10^{-3}	4.95×10^{-3}	219
	20	1.53	2.28	149
	50	1.53	2.45	160
Alanine-U-^{14}C	3	3.39×10^{-3}	17.0×10^{-3}	502
	50	4.87	21.0	432

Average values from 6 to 30 determinations

TABLE 4

Incorporation rate (K_p) of the amino acids lysine, glycine and α-alanine into protein of male adults of <u>D</u>. <u>melanogaster</u> of different ages

Age (days)	Lysine	Glycine	Alanine
	K_p (μMoles/hr/fly x 10^{-4})		
3	2.18	7.79	4.26
20	1.52	4.65	-
50	1.17	3.23	2.68
$\frac{50}{3}$ x 100	53.7	41.5	62.9
	K_p (μMoles/hr/mg Protein x 10^{-3})		
3	2.51	8.18	6.64
20	1.33	3.99	-
50	0.92	3.42	2.58
$\frac{50}{3}$ x 100	36.9	41.8	38.5

Average values of 15 to 24 determinations

TABLE 5

Oxidation rate (K_r) of lysine, glycine and α-alanine in 3- and 50-day-old male adults of <u>D</u>. <u>melanogaster</u> (μMoles/hr/fly x 10^{-5})

Age (days)	Lysine		Glycine		Alanine	
	K_r	%Pool	K_r	%Pool	K_r	%Pool
3	3.35	5.05	106.9	47.3	777	219
50	3.17	6.73	95.4	62.3	659	135
$\frac{50}{3}$ x 100	94.6		89.4		84.8	

Average values of 2-3 determinations

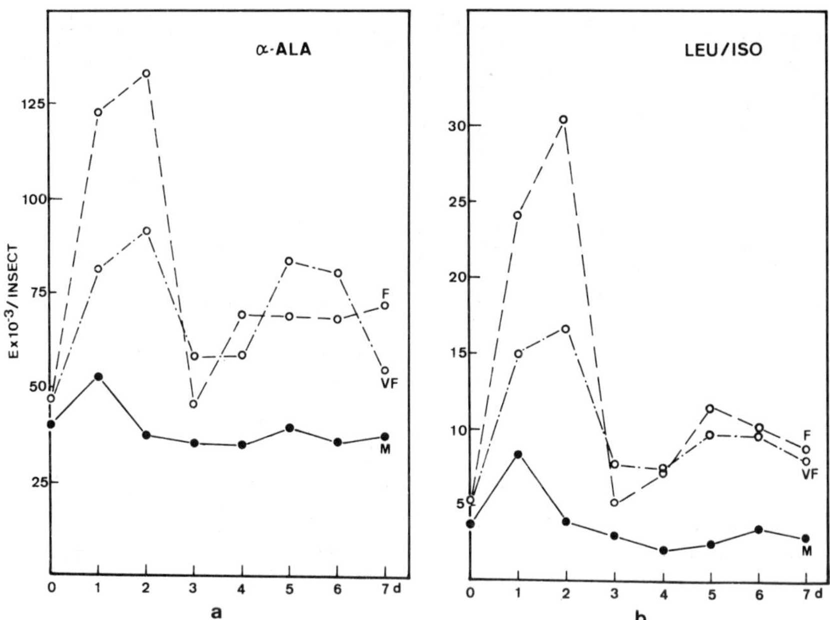

Fig. 1   Changes in the contents of α-alanine (a) and leucine/isoleucine (b) during early adult life of <u>D. melanogaster</u>.   Each point represents the average value of 6 determinations.   Ordinate: extinction (E) per insect.   Abscissa: adult age in days after emergence at 25°C.   F = mated females, VF = virgin females, M = males.

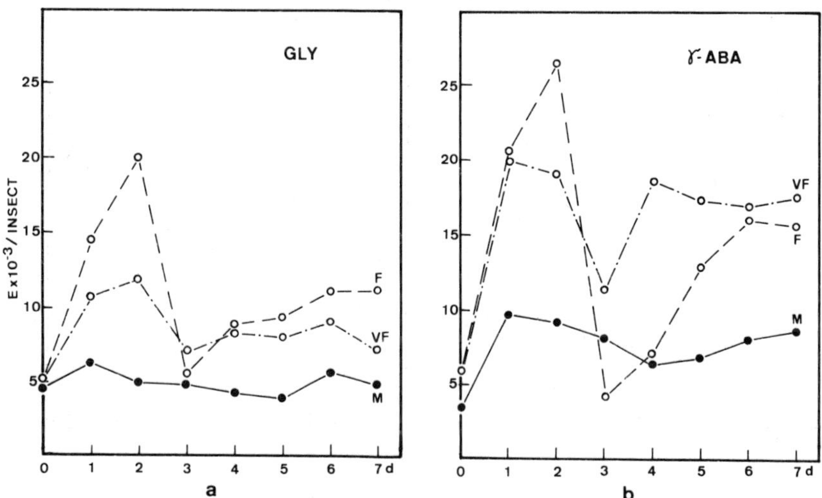

Fig. 2   Changes in the contents of glycine (a) and γ-amino-butyric acid (b) during early adult life of <u>D. melanogaster</u>.   For further explanation, see legend to Fig. 1.

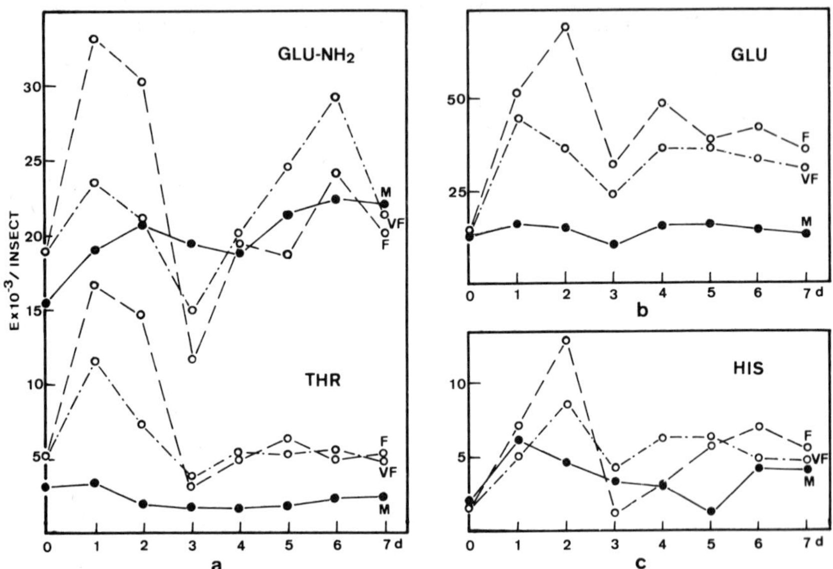

Fig. 3 Changes in the contents of glutamine, threonine (a), glutamic acid (b) and histidine (c) during early adult life of <u>D. melanogaster</u>. For further explanation, see legend to Fig. 1.

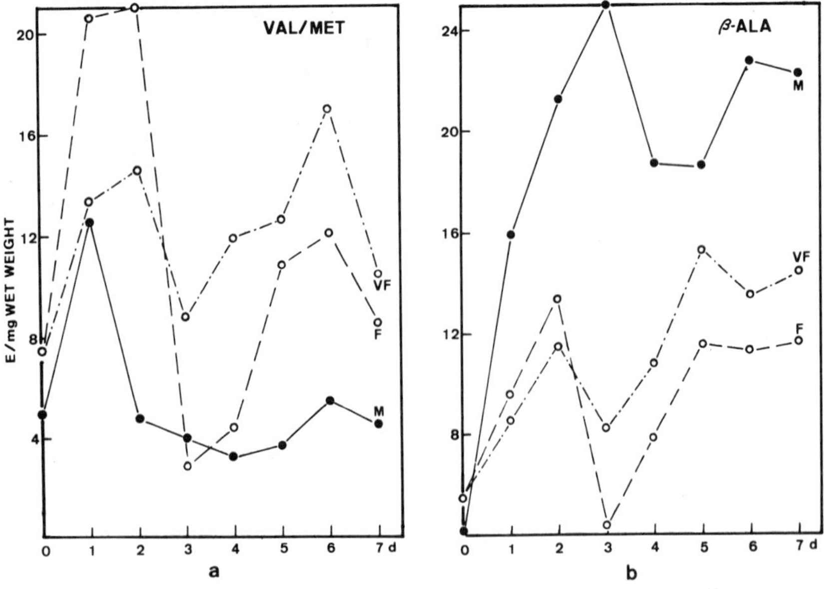

Fig. 4 Changes in the contents of valine/methionine (a) and β-alanine (b) during early adult life of <u>D. melanogaster</u>. Ordinate: extinction per mg wet weight. For further explanation, see legend Fig. 1.

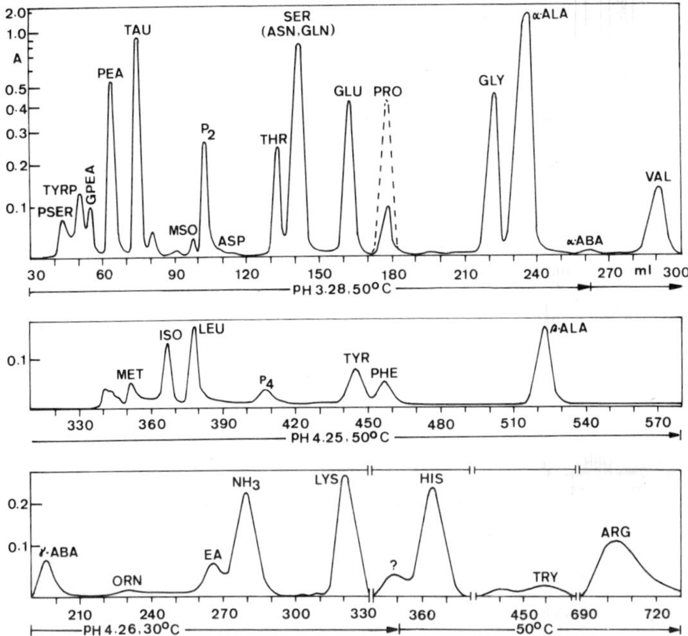

Fig. 5 Chromatographic profile of free ninhydrin-positive components in 0-day-old adult males of
D. melanogaster (0.1 g/1 ml). Ordinate: absorption
(A) at 570 nm (——) and 440 nm (---). Abscissa:
effluent volume in ml. All abbreviations refer to
the first three letters of the corresponding amino
acids with the following exceptions: PSER, phospho-
serine; TYRP, tyrosine-0-phosphate; GPEA, α-glycerophosphoethanolamine; PEA, phosphoethanolamine; MSO,
methionine sulfoxide; ASN, asparagine; GLN, glutamine; ABA, aminobutyric acid; EA, ethanolamine; 3M-
HIS, 3-methylhistidine; PS, paragonial substance; P,
peptide.

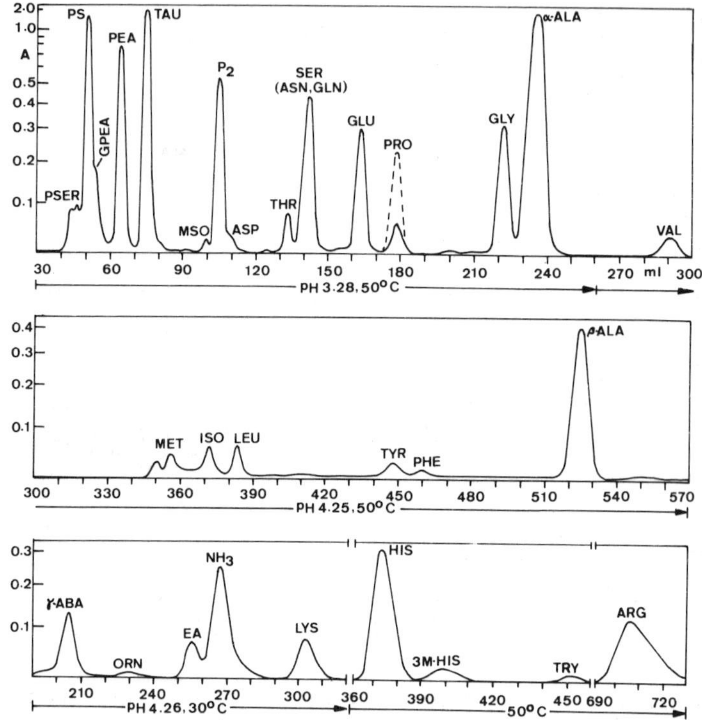

Fig. 6 Chromatographic profile of free ninhydrin-positive components in 30-day-old adult males of <u>D</u>. <u>melanogaster</u>. For abbreviations and further explanation, see legend to Fig. 5.

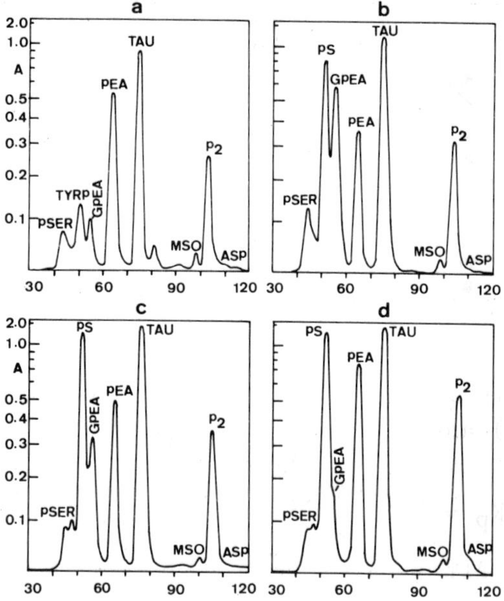

Fig. 7 Changes in the pattern of acidic ninhydrin-positive components in adult males of D. melanogaster aged 0 (a), 3 (b), 20 (c) and 30 (d) days
after emergence. In all diagrams only peaks recorded in the first 120 ml effluent volume of the analyzer are shown. For abbreviations and further explanation, see legend to Fig. 5.

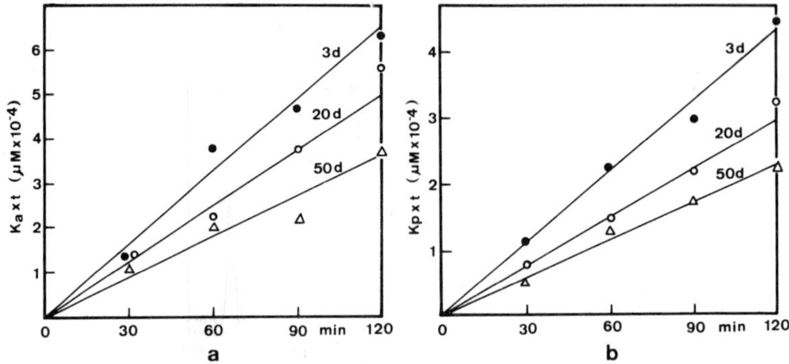

Fig. 8 Turnover (a) and incorporation (b) of free lysine in adult males of $\underline{D.\ \text{melanogaster}}$ aged 3 (●), 20 (o) and 50 (△) days after emergence. Ordinate: turnover rate $K_a = \dfrac{a}{t} \times \ln \dfrac{q(t)}{q(o)}$ ($\mu M \times 10^{-4}$); incorporate $K_p = \dfrac{p(t) \times K_a}{q(o) - q(t)}$ ($\mu M \times 10^{-4}$). Abscissa: time in min. after injection. (From Baumann, 1969.)

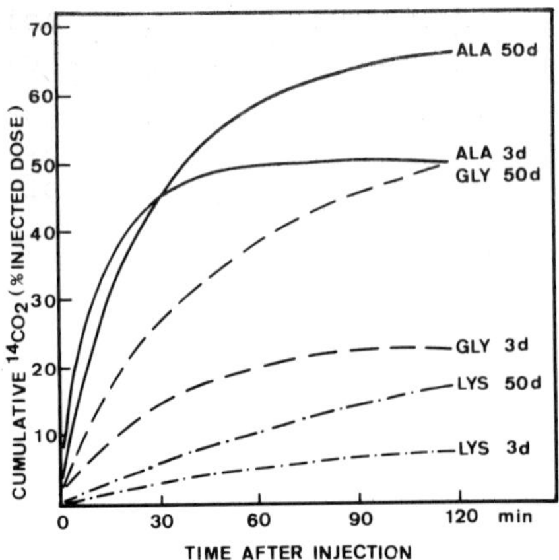

Fig. 9 Cumulative production of $^{14}CO_2$ after injection of uniformly ^{14}C-labeled alanine (——), glycine(---), and lysine (-·-) into adult males of $\underline{D.\ \text{melanogaster}}$ aged 3 and 50 days. (From Baumann, 1969.)

THE OCCURRENCE, INCORPORATION, AND FATE OF
AMINO ACIDS DURING DEVELOPMENT AND AGING IN
THE YELLOW FEVER MOSQUITO, <u>AEDES</u> <u>AEGYPTI</u> (L.)

James D. Stidham, Ph.D.

Department of Biology
Presbyterian College
Clinton, South Carolina 29325

The presence of a high concentration and wide
variety of amino acids in insects is a characteristic feature of this group of organisms. Some insect
species may possess a total concentration of free
amino acids (FAA) of 2 gm per 100 ml, or more than
30 times higher than in other groups of animals.
Studies suggest that the concentration of FAA in
insects is not under direct metabolic control, but
rather varies with the nutritional state, age, and
metabolic activities of the different tissues
(Gilmour, 1965).

Within the past decade we have seen many papers
and several reviews written about the FAA in insects.
In spite of this, however, some very basic questions
concerning the specific functions and behavior of the
complex amino acid pools during development and aging
have remained unanswered. Some limitations of past
studies have included the lack of comparability of
the various studies due to differences in techniques,
the measuring of FAA titers in hemolymph only, lack
of knowledge of the exact physiological age (Stephen
and Steinhauer, 1963), differing diets and environmental variations.

Thus it should be evident that in any study of
FAA changes in insects during development and aging,
attempts should be made to control as many of the
above variables as possible. The mosquito larva,

pupa and adult of <u>Aedes</u> <u>aegypti</u> are very well suited
for this type of study since their environment and
diet can be accurately controlled, and thereby their
physiological age.

Aside from such special cases as the formation
of the cocoon and synthesis of cuticle, the physiological significance of the high AA titre still
remains obscure. The long-considered function of
this AA pool has been as a storage reservoir for use
in protein synthesis. Chen (1966) has noted, however, that the AA greatly exceed requirements for
protein synthesis, thus suggesting other functions
of perhaps equal importance. Since the titre of FAA
seems to be considerably higher in Holometabola than
in Hemimetabola, it has been suggested that they
might serve a special function in metamorphosis
(Levenbook, 1966). The use of this pool as an
energy reserve has also been implicated by several
workers (Bursell, 1966; Mansingh, 1966; Stidham,
1969).

A considerable body of information regarding
protein synthesis during insect development and
aging is rapidly becoming available. The procedures
commonly employed for such studies are the injection
of labeled AA into the young or adults, or allowing
the insects to feed on a medium containing the
labeled materials.

Few studies have investigated the total utilization of the FAA via oxidation, excretion, and
tissue incorporation during development and aging
of insects. Our studies with <u>Aedes</u> <u>aegypti</u> have
been devoted to the investigation of developmental
changes in the free and peptide amino acids during
the larval and pupal stages as well as in the aging
adult female mosquito. A second aspect of our study
was undertaken to determine the metabolic interrelationships occurring during development and aging
in this insect.

Changes in free amino acids and peptide amino
acids (PAA) during the third and fourth larval instars and during the pupal stage were investigated

by the method of automatic ion-exchange chromatography (Technicon Auto-Analyzer). AA were precipi-
tated by the use of cold 80% ethanol. Portions from
each sample were hydrolyzed to free the peptide-
bound AA. A total of 36 ninhydrin positive peaks
were initially found. The 18 AA consistently present
were: methionine sulfoxide, aspartic acid, threonine,
serine, glutamic acid, proline, glycine, alanine,
valine, cystine, methionine, isoleucine, leucine,
tyrosine, phenylalanine, lysine, histidine, and
arginine. Ornithine, taurine, urea, ethanolamine,
and the α, β, γ, forms of amino-butyric acid were
also present. Peptide AA values were obtained by
determining the differences in AA concentration
between hydrolyzed and unhydrolyzed samples (Chaput,
1969).

The values are expressed in μmoles of AA/ml of
body water. Since conventional biochemical practice
is to express data on the basis of wet or dry weight,
a brief explanation of our usage is necessary here.
Table 1 shows the observed changes that occurred in
wet and dry weight as well as in total body water
during the larval and pupal stages of <u>A</u>. <u>aegypti</u>.
The wet:dry weight ratio is also given. The drop in
wet weight at pupation is apparently due to a
corresponding drop in water volume, since the dry
weight actually increases at this time. Therefore,
if the FAA concentration were calculated on basis of
wet weight, the concentration values obtained would
increase at pupation, whereas if calculated on the
basis of dry weight, the concentration values obtained would decrease during pupation. It becomes
evident, then, that an accurate FAA concentration
can only be obtained on the basis of ml of body
water. We shall here refer, not to every AA, but
rather to certain ones which are either representative of a group of AA or show unusual variations.

The concentration of free phenylalanine remained comparatively constant during the developmental period studied. Although the concentration
of phenylalanine in peptides was relatively low,

the higher values at the end of each instar and the
pupal stage may reflect developmental processes.
Fig. 1 exhibits the profiles for phenylalanine. A
rise in the FAA titres following pupation was fairly
representative of most of the FAA. The fall in
peptide amino acids at pupation, followed by a subsequent rise during the pupal stage, was also characteristic of nearly all of the AA isolated. Isoleucine, leucine and valine exhibited basically the
same concentration profiles (Fig. 1). The free and
peptide AA concentrations showed minimal variations
in the larval instars. The characteristic increase
in concentration of FAA and decrease in PAA at
pupation and the rise in PAA during the pupal stage
were evident.

The concentration of free proline (Fig. 1) rose
to a maximal value during the first half of the
third and fourth instars and fell to minimal values
during the latter half of the instars. The concentration remained fairly constant during the pupal
stage. The low concentration in peptides was
comparatively constant during larval development.
The typical decrease at pupation and increase during
the pupal stage were present.

The FAA titre of aspartic acid (Fig. 2) was
consistently low and exhibited minimal variations.
The peptide level of this AA was relatively higher
than any other amino acid. Its concentration showed
a downward trend during the larval instars followed
by a rapid rise to initial levels during the pupal
stage. Threonine showed a high FAA concentration in
early instars but dropped toward the pupal stage
with a very low concentration in PAA.

Although the FAA level of alanine (Fig. 3)
exceeded all other AA, it had relatively the smallest
PAA titre. The concentration of free alanine followed a downward trend during the development period
observed. Glycine also had a relatively high PAA
titre in contrast to its FAA titre. Both free and
peptide glycine showed minimal changes during
development (Fig. 3). Free glutamic acid changed

little during the developmental period observed.
The drop in peptide-bound levels at pupation was not
followed by the anticipated increase during the pupal
stage (Fig. 3).

The amino acids arginine, histidine, and tyrosine all exhibited the same profiles during development (Fig. 2). Their FAA levels showed a gradual
decrease during the larval development, followed by
an increase during the pupal stage.

In looking at changes in the <u>total</u> concentration
of free and peptide-bound amino acids (Fig. 4), we
observe a drop in total FAA concentrations during
development, paralleled by an equivalent drop in the
concentration of non-essential FAA. The essential
amino acid concentration, however, remained very
constant throughout the same developmental period.
The peptide-bound AA concentration for both the
essential and non-essential AA remained rather uni-
form. The only variation occurred at pupation where
the level fell sharply only to rise to initial levels
during the pupal stage. Recall here that this latter
characteristic was found to occur in nearly all of
the PAA observed.

In summary, free alanine was the most prominent
of the total free amino acid pool, comprising between 12 and 18 percent. Over 50% of the peptide
amino acids were contributed by the three amino
acids, glutamic acid, aspartic acid, and glycine.

When changes in amino acid composition had been
determined for the larval and pupal stages, studies
on AA metabolism during this same period of development were initiated. These employed C-14 labeled
amino acids and were designed primarily to determine
the utilization rates of several of the amino acid
pools. The isotopes were used extensively as a
means of following the metabolism of individual AA
and of the pools as a whole. The time of exposure
to the isotopes was set at 70 minutes since at this
time the level of AA radioactivity was relatively
high compared to the protein level.

Third and fourth instar larvae reared asepti-

cally were allowed to feed on media containing one
of the following uniformly labeled C-14 amino acids,
glutamic acid, aspartic acid, alanine, or phenylalanine. Pupae were injected with these labeled compounds with histidine substituted for phenylalanine
because the latter proved lethal to pupae. The distribution of radioactivity in the amino acid and protein fractions were investigated to determine the
rates of amino acid utilization and protein synthesis. Fig. 4a shows the distribution for glutamic
acid uniformly labeled with C-14 in the FAA and proteins of the third larval instar. All other AA
studied followed similar distribution profiles. The
uptake of the isotope was quite rapid during the
feeding period and then decreased slowly after removal of the larvae from the labeled media. The incorporation of label into the protein fraction was
delayed. No lipid fractions were analyzed since
preliminary studies had shown an insignificant amount
of activity incorporated into this fraction two
hours after feeding on a labeled diet.

As a criterion of the oxidation of the labeled
amino acids, $^{14}CO_2$ production was measured from both
larvae and pupae (Figs. 5a, 5b). As might be expected, the highest oxidation rates were found very
shortly after initiation of feeding or after injection of the label (pupae). It should be evident
that, at corresponding stages of development, the
non-essential amino acids aspartic acid, glutamic
acid, and alanine were more rapidly and extensively
oxidized than were the essential amino acids histidine and phenylalanine. Aspartic acid oxidation was
more extensive (65-80%) than any other AA during all
stages of development, except for the post-molt
pupae, where glutamic acid oxidation was slightly
higher than aspartic acid. The other amino acids
were oxidized in order of extensiveness as follows:
glutamic acid, alanine, histidine and phenylalanine.
The differences in oxidation relative to developmental period seemed to be minimal. In general, all
of the AA studied in the pupal stage showed an

increase in oxidation rate from middle to pre-molt pupae (Fig. 5b).

In addition to determining the oxidation rates for the various AA, the loss of radioactivity through excretion was also estimated. By determining the specific activity and total volume of the non-labeled medium after the last CO_2 sample had been obtained and knowing the number of larvae in each culture, the activity excreted per larva was calculated. The results, expressed as a percent of the total activity ingested, indicated a smaller proportion of activity was excreted when using isotopes of non-essential AA than when using isotopes of essential AA (e.g., 3.9% for aspartic, 4.7% for alanine and 2.0% for glutamic acid, compared to 14.1% for phenylalanine and 24.1% for histidine).

If we now turn our attention to utilization rates for the different amino acids, we see these to be markedly different. The absolute rate for alanine in the pre-molt fourth instar larvae was over 5 times greater than glutamic acid and over 50 times greater than aspartic acid and phenylalanine. Except for the lower value in pre-molt third instar larvae (where the utilization rate dropped below that of glutamic acid), the rate of alanine utilization remained consistently higher than all of the other AA studied. Another interesting point here is that the non-essential amino acids were utilized at a faster rate than the essential AA. All the utilization rates showed a general increase during larval development with maximum rates occurring at the end of the fourth instar. The rates decreased during the first half of the pupal stage and then rose again during the last half of the pupal stage.

In order to compare the extent to which the individual AA pools were being utilized, the percent of AA utilized was plotted against developmental stage. These data are seen in Fig. 6. The curves here indicate quite clearly that the individual pools were used at rates which seem to be independent, to some degree, from one another. It is interesting to

note that pool utilization is lower in the two
essential amino acids, phenylalanine and histidine.
We should also note here that the utilization of the
individual AA during metamorphosis (data presented
in Fig. 6) follows, to some degree, a U-shaped profile which is characteristic of insect respiration
during this period of development.

The amount of radioactivity found in the protein fraction of larvae after having fed for 70 min.
on labeled media and in the protein fraction of
pupae 60 min. after having received an isotope injection were arbitrarily chosen for comparing the
rates of individual AA incorporation into proteins
at different times during larval and pupal development. The larval values were corrected for excretory
losses and recomputed from cpm per 100 larvae to
percent of total activity ingested. The resulting
data are shown in Fig. 7.

It should be evident here that the AA were
incorporated into proteins at essentially different
rates. Phenylalanine exhibited the highest rate of
incorporation followed by alanine, histidine,
glutamic acid and aspartic acid in that order. However, all of the AA studied exhibited similar incorporation profiles during development. Thus, the
highest rates of incorporation occurred in early
third instar followed by a general decrease to low
values in the fourth instar. Thereafter, the rates
increased, phenylalanine showing the most marked
rise. The rates of incorporation remained quite
constant during the pupal stage.

At this point, let us look briefly at the FAA
composition and resultant changes during aging in
the adult.

Adult mosquitoes were given two different
dietary regimens (sucrose, and sucrose and blood)
and sacrificed at seven different ages from 6 days
after emergence to 45 days. Sixteen FAA were consistently present; these were alanine, aspartic acid,
glutamic acid, glycine, serine, threonine, proline,
valine, methionine, isoleucine, leucine, tyrosine,

phenylalanine, lysine, histidine and arginine. Four
other AA were also present but in rather low concentration: taurine, cysteic acid, methionine sulphoxide and Beta-alanine. The concentration of FAA
in those mosquitoes given blood and sucrose was
consistently higher than in individuals given only
sucrose. Fig. 8b shows the total FAA concentration
in both groups. Evident here is the fact that the
total AA level is quite high the first week (9 days)
in blood-fed individuals and drops to a lower concentration the second week (16 days) before rising
substantially to a peak at the fourth week (30 days).
However, the total AA concentration in sucrose-fed
mosquitoes, by comparison, is rather low during the
first week and a half (i.e., at 6 and 9 days) and
increases after the second week (16 days), reaching
a peak after the end of the fourth week (30 days).

Alanine was the FAA most abundant in both experimental groups, comprising from 34 to 52% of the
total FAA load. A comparison of total FAA concentration and the concentration of alanine alone is
depicted in Fig. 8a. Values here are expressed in
μM/mg sample dry weight. Proline, aspartic acid
and the basic amino acid arginine were present in
rather high concentrations when compared with the
remaining FAA. These findings are in agreement
with most published reports concerning FAA in other
invertebrates and the predominance of one or two
acidic and/or neutral AA and at least one basic AA
(Wyatt, 1961). In the family Culicidae to which all
true mosquitoes belong, alanine appears to be the
FAA in greatest concentration (Chen, 1963; Duffy,
1964; Stidham, 1968). It should not seem surprising
to find a non-essential AA, or one that can be
readily synthesized by the insect, occupying the
central role in the nitrogen metabolism of an
organism. Such an amino acid could be synthesized
by several pathways and would also make up a large
percentage of many peptides in the system. Chen
(1963) hydrolyzed a number of peptides from the
female mosquito, _Culex_, and found several non-

essential AA to predominate: alanine, aspartic acid,
glutamic acid, glycine and proline.

Another possible role of alanine could be in
energy metabolism. Winteringham (1958) has suggested that, in insects, some of the AA may provide a
soluble and readily available substrate reserve for
the Krebs' cycle. Bursell (1963, 1966) and Chen
(1966) give evidence that alanine and proline can be
used by some insects as energy reserves. Bursell
(1966) has shown that after injection of radioactive
alanine about 75% of the activity could be assigned
to specific AA (proline, alanine, glutamic acid),
the rest appearing as organic acids of the Krebs'
cycle.

Price (1961) has suggested that alanine could
well be a glycolytic end-product formed from pyruvate
by amination.

Alanine and/or aspartic acid may play the central role in metabolism in some lower animals, a role
normally reserved for glutamic acid in mammalian
metabolism. That alanine does indeed play the central role in metabolism in the adult _Aedes aegypti_
would not be surprising in view of its tremendous
concentration both in sucrose- and sucrose-and-
blood-fed mosquitoes (Fig. 8b).

All of this seems to lend support to our
suggestion that alanine occupies the central role in
metabolism in adult female _Aedes aegypti_, that it
exists in a stable relationship to the other amino
acids, and that it probably serves as the "carbo-
hydrate-amino acid link" in this insect.

In an effort to learn more about the role and
metabolic fate of amino acids in the adult, we investigated the metabolism of certain labeled amino
acids; compounds used were the uniformly labeled
C-14 isotopes of alanine, aspartic acid and isoleucine. The first two are non-essential and the
last, essential. The materials and methods, together with details of this work, are available in
a previous report (Stidham and Liles, 1969).

After feeding labeled alanine, the activity of

the AA fraction reached 2600 cpm at 12 hr, then declined to a low of 32 cpm at 42 days. Fig. 9b shows these data for the three fractions, viz., amino acids, proteins and lipids. Here we see an abrupt drop in activity between 12 hr and 48 hr post-feeding. While it is apparent here that the activity at 2 days is higher than at 42 days, variations in activity between day 2 and day 42 were found to be statistically insignificant.

Radioactivity in the lipid fraction was highest at 2 days (1800 cpm). A gradual decline then followed to a low at 42 days (~600 cpm per mosquito).

Changes in radioactivity of the protein fraction over the 42-day period were initially high with approximately 220 cpm, rising to above 300 cpm. Activity then decreased sharply to day 6 and then more slowly thereafter to 42 days.

Analysis of the AA fraction from mosquitoes fed labeled aspartic acid revealed an initial value of 600 cpm/mosquito at 12 hr after feeding. This value declined during the next 2 weeks to about 24 cpm/mosquito at 15 days post-feeding. Radioactivity of three tissue fractions from mosquitoes fed labeled aspartic acid and analyzed over two weeks is revealed in Fig. 9a. The 12-hr value was statistically different from all remaining values. The remaining values for the 2-week period were not statistically significant from each other.

The lipid fraction showed maximum activity at 12 hr after feeding, followed by a sharp decline to day 1. Activity then increased gradually to day 11 and declined to day 15. Radioactivity of the protein fraction was highest at 12 hr after feeding and then declined gradually during the remaining 2-week period. The values at days 1 and 3 were somewhat lower than the remaining values and were found to be statistically different.

The data for radioactivity of three tissue fractions from mosquitoes fed C-14 labeled isoleucine are shown in Fig. 10 for both egg-laying and non-egg-laying females. Examination of the AA fraction data

of the non-egg-producing group revealed a rapid uptake, the high point, then a rapid and significant
decrease over the first 24 hrs following removal of
isotope. After this initial decrease, a further
decrease was observed during the remainder of six
weeks. Although both the egg-producers and the non-
egg-producing group tended to lose radioactivity,
the non-egg-producers did so at a higher rate; could
the latter group be retaining same for future egg
production?

As the activity of the AA fraction decreased,
the activity of the lipid and protein fraction was
seen to increase. Again, the initial rate of incorporation appeared slightly greater in the non-
egg-producers, but much more lipid appears to be
synthesized from the isoleucine by the egg-producers.
Examination of the lipid data revealed no statistical difference between the initial time and any of
the remaining times, for the non-egg-producers.
However, a trend of increase over the 6-week period
for the non-egg-producers was observed. Data for
the egg-producers showed an initial increase,
reaching a maximum at about 2 weeks, and then de-
creasing until the end of the experimental period.

Protein synthesis in both groups greatly exceeded lipid incorporation. In both groups, activity
reached a peak and then decreased during the remainder of the experimental period. Data from the
non-egg-producing group would suggest this peak was
reached during the first 3 days. The maximum values
of activity of both groups were nearly equal, indicating a rather similar ingestion of isoleucine by
both groups.

The decrease in the AA fraction closely paralleled the increase of the protein fraction in both
groups. Most of the incorporation by the non-egg-
producers appeared to occur during the first 36 hrs.
While the incorporation by the egg-producing group
also was greatest during the initial 36 hrs, it
seems to have continued, to some extent, during the
remainder of the first week. There is also the

suggestion of renewed protein incorporation by the egg-producers between 35 and 42 days. Perhaps significant here is the finding by Liles (1965) that mated females consistently live longer than virgin females (about 17% longer). This worker feels that the hormone cycles in the mated females initiated by the mating process are beneficial in some, as yet, unknown manner. Rockstein et al. (1971) showed that, if female house flies lay several batches of eggs during the course of their adult life span, they live significantly longer than those individuals which are allowed to lay only one or two batches of eggs. They state that these flies release more hormones during their adult life than do flies which do not lay eggs, and, therefore, that each time the hormone is released certain enzymes could be theoretically reprogrammed. Moreover, these workers suggest that, based on histological evidence, one might theorize that failure of the median neuro-secretory cells to secrete hormones is producing disturbances in protein and carbohydrate metabolism. This could then account for the reduction in synthe-sis of specific enzymes, which, in turn, would result further in the failure to mobilize those nutrients necessary to maintain homostasis in the aging insects.

In another study, Tietjen (1967) found that amino acids obtained prior to the production of the first batch of eggs were still available for use in the production of the second batch of eggs. His data also showed that amino acids placed by the female into the F_1 eggs could be used in the pro-duction of the F_2 eggs by their daughters. There appeared to be little change in the protein in various larval stages, but data did suggest that some lipid was disappearing (for energy production?). In view of the known histolytic processes associated with molting, it is of interest that the percent protein showed very little change between the larval stages and the adult. In an effort to gain informa-tion about the movement of the C-14 obtained from

the larval diet into eggs, Tietjen then analyzed
the resulting adults.

Data for the three extract fractions showed a
gradual loss over a 3-week period. Unfortunately,
the few individuals available for this study made
it impossible to use proper replication, hence we
can only observe trends. The data suggest an incorporation of the amino acid into a relatively
stable protein at the time of pupation.

In order to further establish the metabolic
fate of the labeled compounds, we sought to determine
whether mosquitoes given labeled amino acids would
expire labeled CO_2. The production of labeled CO_2
by mosquitoes given labeled alanine and labeled
aspartic acid is shown in Fig. 11a. There was rapid
loss of $^{14}CO_2$ from mosquitoes regardless of the
labeled AA they had received. After 21 hr post-
feeding, there were no significant changes in $^{14}CO_2$
production. These data suggest a very rapid release
of label via this route during the early hours after
uptake followed by a somewhat constant level of
production for several hours. The rates of the loss
through CO_2 at various ages from mosquitoes given
labeled isoleucine is seen in Fig. 11b. This rate
of loss was much lower than for the non-essential
amino acids, alanine and aspartic acid. Non-egg-
producers showed an increase in loss for the first
20 hrs, then a gradual loss. Egg-producers showed
greater loss for the first 36 hrs followed by a low
rate of loss similar to the loss of non-egg-producers. It cannot be said with certainty whether
the oxidation of the amino acids was direct (via its
intermediary metabolic route) as Bursell (1963)
suggests for proline in the tsetse fly, or whether
the carbon is first converted into lipid, or carbohydrate and then oxidized.

Perhaps the last avenue to be analyzed as a
possible source of loss of carbon-14 was mosquito
excreta. Analyses were made at 6-hr intervals from
6 hr after uptake of labeled AA through 120 hr.
Data collected from analysis of excreta deposited on

the bottoms of adult rearing cages were subjected to
statistical procedures. No statistical differences
were found among any values from either labeled
alanine or aspartic acid. There was, however, some
small loss, as counts for the total excreta matter
were at least 35 counts/min above background. This
loss was quite small for these two non-essential AA
but did represent a very real loss by this route.

Loss of activity through excreta from adult
mosquitoes fed labeled isoleucine showed the non-egg-
producers lost most of activity by the end of the
first 24 hrs. Egg-producers were seen to lose most
of activity between 36 and 48 hrs following ingestion
of the labeled isoleucine. The loss values of the
non-egg-producers were consistently higher than the
comparable values of the egg-producers.

ACKNOWLEDGEMENTS

This manuscript was completed while the author
held a NSF Science Faculty Fellowship tenured at the
School of Marine and Atmospheric Science, University
of Miami. Grateful acknowledgement is extended to
the Division of Functional Biology for the use of
facilities and aid in preparing this report.

The author extends deep appreciation to Dr.
James Liles, Dr. W. Tietjen, and Dr. R. Chaput for
the use of portions of their data, and for many
helpful discussions of the research herein reported.

This research was supported in part by a
contract (No. DA-18-064-48) from the United States
Army Biological Laboratories, Fort Detrick, Fredrick,
Maryland.

REFERENCES

Bursell, E. (1963). J. Insect Physiol. 9, 439.
Bursell, E. (1966). Comp. Biochem. Physiol. 19, 809.
Chen, P.S. (1963). J. Insect Physiol. 9, 453.
Chen, P.S. (1966). In "Advances in Insect Physiology"
   (J.W.L. Beament, J.E. Treherne and V.B. Wiggles-

worth, eds.), Vol. III, pp. 53-132. Academic
Press, New York.

Chaput, R.L. (1969). *Ann. Entomol. Soc. Amer.* **62**,
742.

Duffy, J.P. (1964). *Ann. Entomol. Soc. Amer.* **57**, 24.

Gilmour, D. (1965). "The Metabolism of Insects."
W.H. Freeman Co., San Francisco.

Levenbook, L. (1966). *Comp. Biochem. Physiol.* **18**,
341.

Liles, J.N. (1965). *Mosquito News* **25**, 435.

Mansingh, A. (1966). *J. Econ. Entomol.* **59**, 234.

Price, G.M. (1961). *Biochem. J.* **80**, 420.

Rockstein, M., Gray, F.H., and Berberian, P.A. (1971).
Exp. Geront. **6**, 211.

Stephen, W.P., and Steinhauer, A.L. (1963). *Proc.
Entomol. Soc. Wash.* **65**, 99.

Stidham, J.D., and Liles, J.N. (1968). *J. Insect
Physiol.* **15**, 1969.

Stidham, J.D., and Liles, J.N. (1969). *Comp. Biochem.
Physiol.* **31**, 513.

Tietjen, W.L. (1967). Ph.D. Thesis, University of
Tennessee, Knoxville.

Winteringham, F.P.W. (1958). *Proc. Int. Symp.
Microchem., Birmingham.* 305.

Wyatt, G.R. (1961). *Ann. Rev. Entomol.* **6**, 75.

TABLE I

THE OBSERVED WET, AND DRY WEIGHT AND WATER VOLUME
OF LARVAE AND PUPAE <u>AEDES</u> <u>AEGYPTI</u> (L.)[a]

Developmental stage	Wet wt/ insect (mg)	Dry wt/ insect (mg)	Ml of[b] water/ insect (10^{-3})	Dry:Wet wt ratio
Third Instar				
Post-molt	0.620	0.070	0.550	0.113
Middle	1.330	0.203	1.127	0.153
Pre-molt	1.771	0.315	1.456	0.178
Fourth Instar				
Post-molt	2.992	0.521	2.471	0.174
Middle	3.673	0.709	2.964	0.193
Pre-molt for males	4.112	0.802	3.310	0.194
Pre-molt for females	5.028	0.975	4.053	0.194
Pupal Stage				
Post-molt	4.604	1.060	3.544	0.230
Middle	4.893	1.223	3.670	0.250
Pre-molt	5.100	1.275	3.825	0.250

[a]All weights were determined from a sample of at least 100 larvae or pupae.
[b]Calculated as (wet-dry) weights. Since water has a specific gravity of approximately one, weight is equivalent to volume.

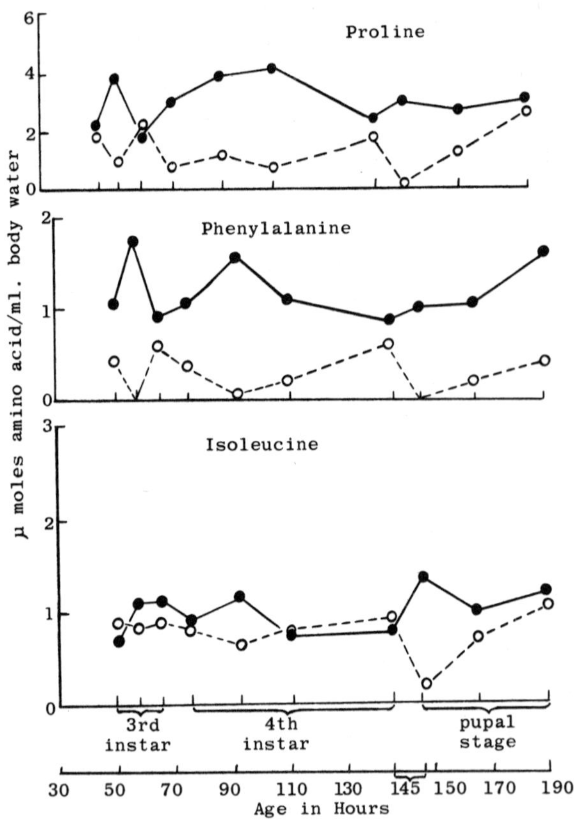

Fig. 1   Changes in free and peptide proline, phenylalanine, and isoleucine concentrations during development.   (FAA ●—●; PAA o—o)

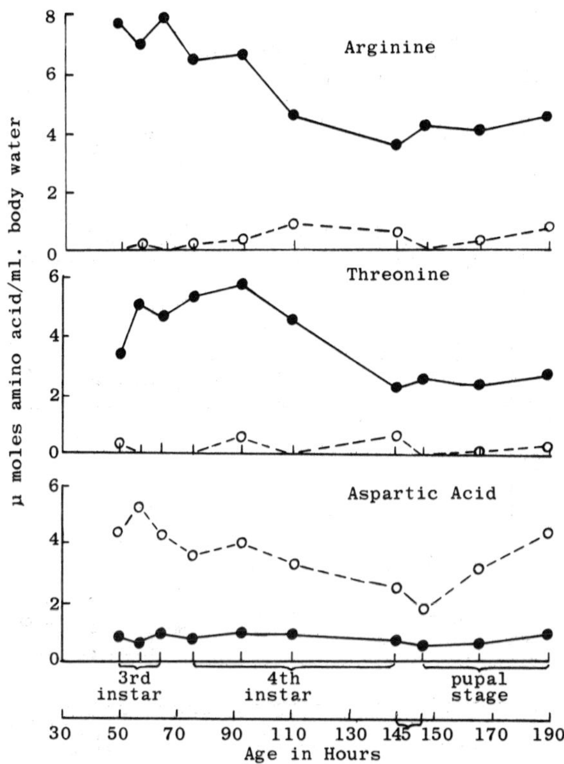

Fig. 2 Changes in free and peptide arginine, aspartic acid, and threonine concentrations during development. (FAA ●—●; PAA o—o)

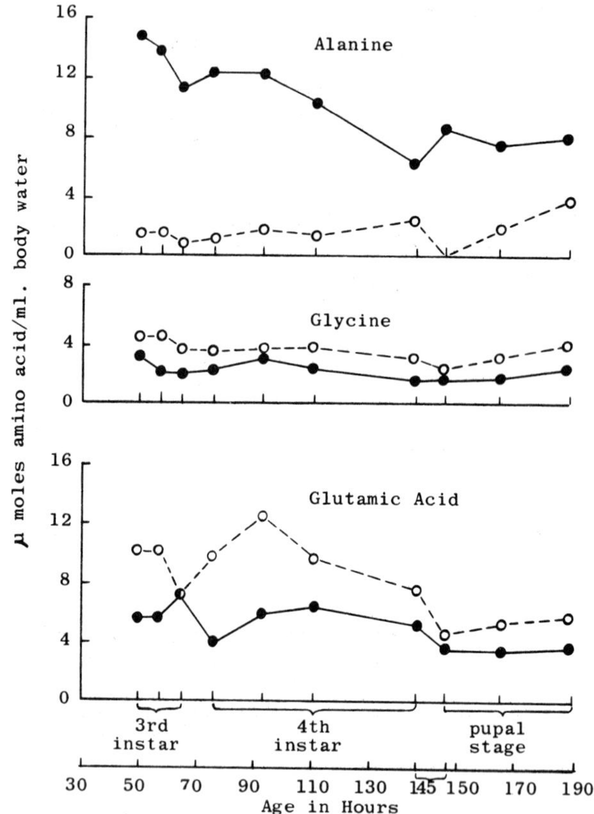

Fig. 3 Changes in free and peptide alanine, glycine, and glutamic acid concentrations during development. (FAA ●—●; PAA o—o)

246

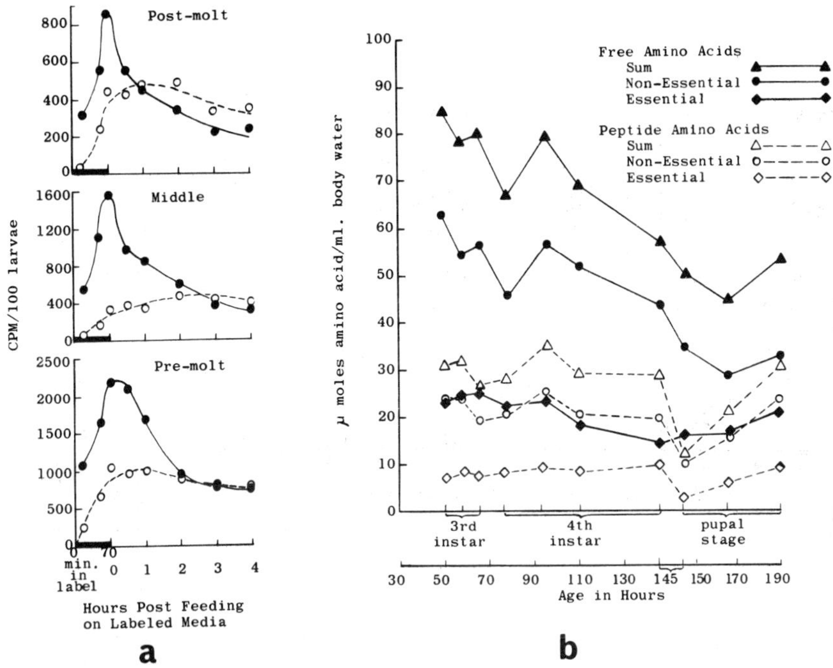

Fig. 4a   Incorporation of radioactivity by third instar larvae from glutamic acid -U-^{14}C.   (amino acids ●—●; proteins o—o)

Fig. 4b   Changes in total free amino acid and peptide amino acid concentrations during development.

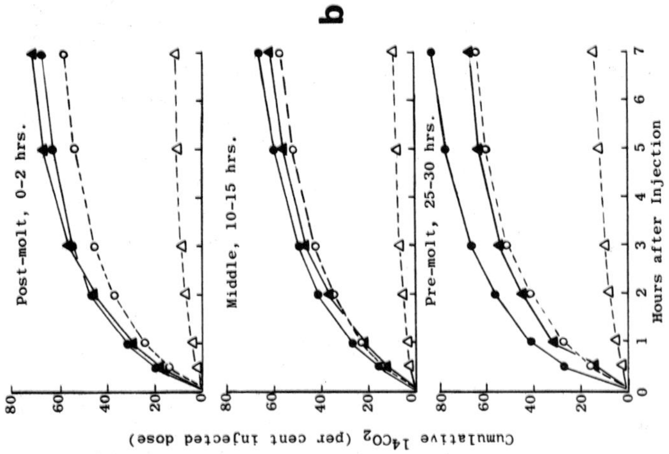

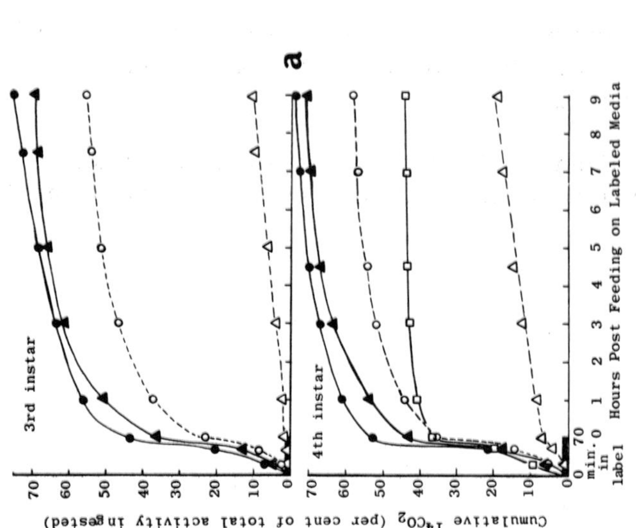

Fig. 5 Oxidation of amino acids during development. a. By third and fourth instar larvae. b. By pupae of various ages. (aspartic acid ●—●; glutamic acid ▲—▲; alanine o—o; phenylalanine △—△; histidine □—□)

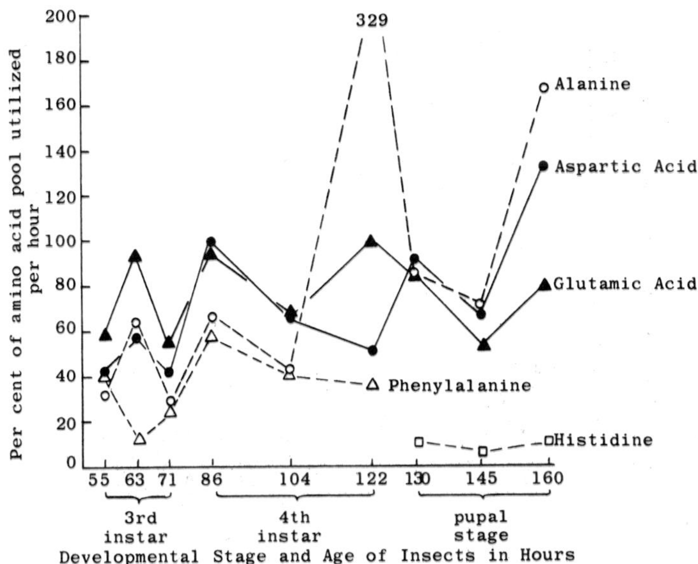

Fig. 6 Utilization of several amino acid pools during development.

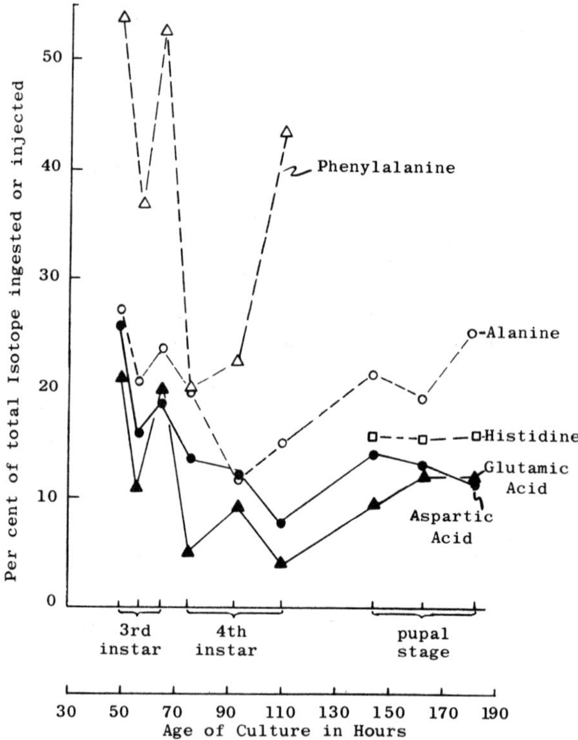

Fig. 7 Radioactivity of larval and pupal proteins - rate of incorporation into proteins of several amino acids (percent of total isotope ingested or injected).

250

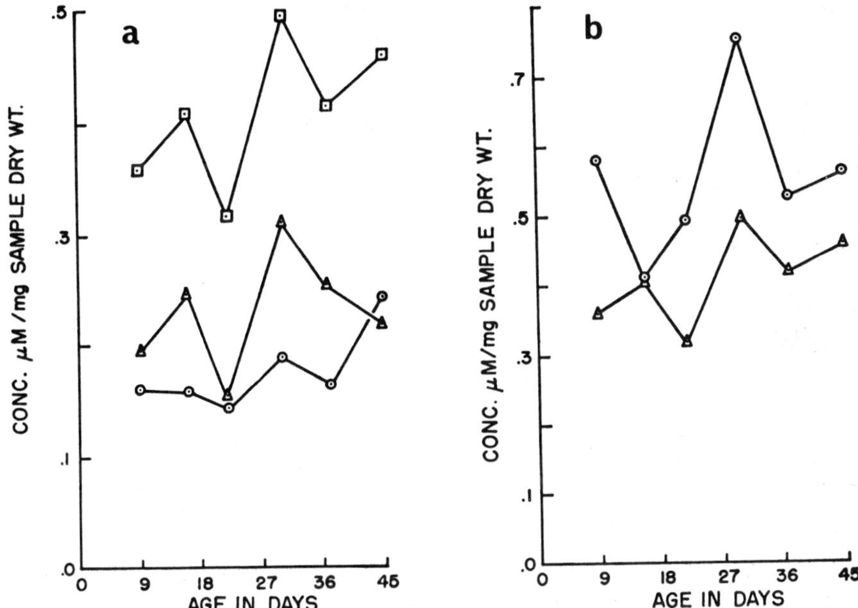

Fig. 8a   Comparison of total free amino acid
concentration and free alanine concentration in adult
female mosquitoes. (□ total amino acid concentration; △ concentration of all amino acids minus
alanine; o concentration of alanine.) Age after
emergence.

Fig. 8b   Total free amino acid concentration
in adult mosquitoes from two dietary regimens.
(△ sucrose fed; o sucrose and blood fed)

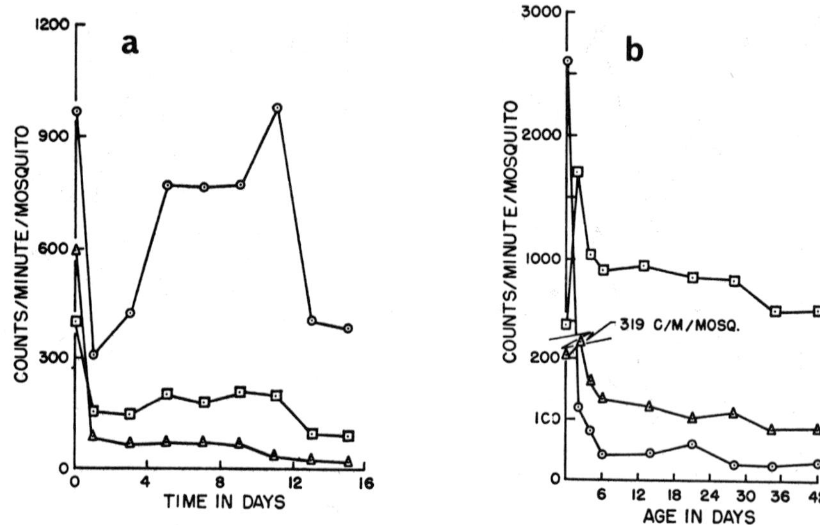

Fig. 9a   Radioactivity of three tissue fractions from adult mosquitoes given aspartic acid -U-^{14}C. (o lipid fraction; $\triangle$ amino acid fraction; $\square$ protein fraction)

Fig. 9b   Radioactivity of three tissue fractions from adult mosquitoes given alanine -U-^{14}C. ($\square$ lipid fraction; o amino acid fraction; $\triangle$ protein fraction)

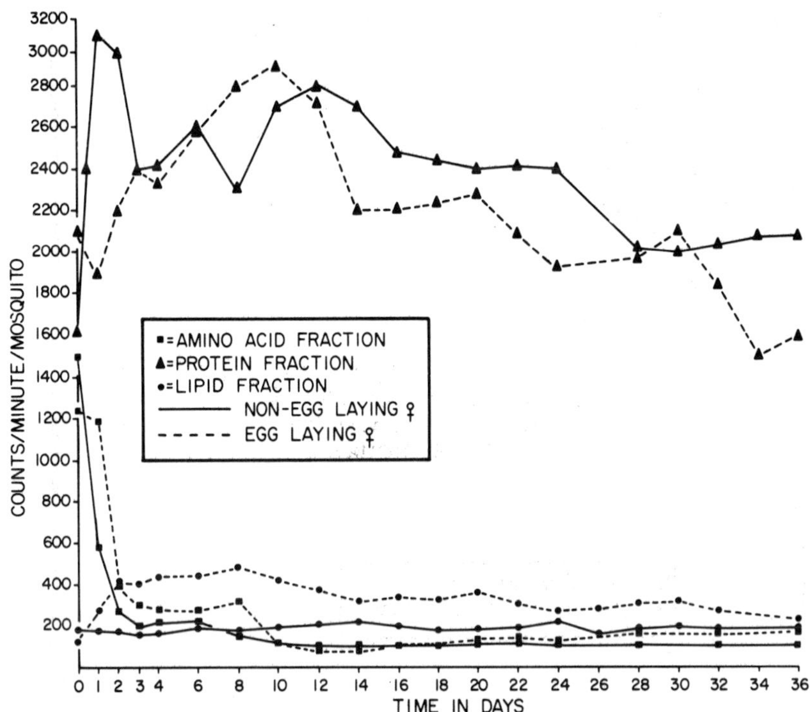

Fig. 10 Radioactivity of three tissue fractions from adult mosquitoes given isoleucine -U-^{14}C.

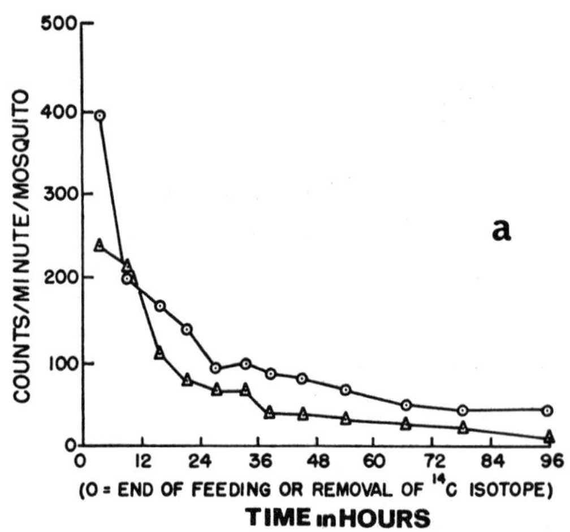

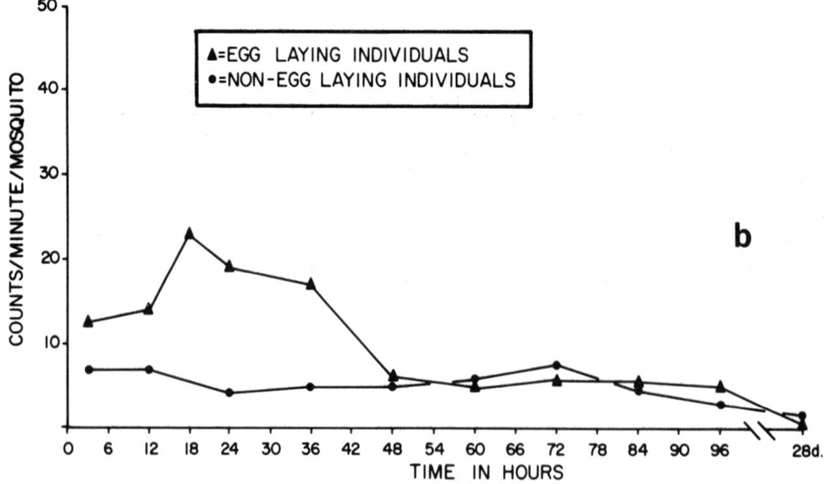

Fig. 11 Production of $^{14}CO_2$ by adult mosquitoes. a. $^{14}CO_2$ from mosquitoes fed alanine -U-^{14}C (o), and aspartic acid -U-^{14}C (Δ). b. $^{14}CO_2$ from mosquitoes fed isoleucine -U-^{14}C.